项目案例开发丛书

jQuery+Vue.js+Spring Boot 贯穿式
项目实战 微课视频版

陈冈◎编著

清华大学出版社
北京

内 容 简 介

jQuery、Vue.js 和 Spring Boot 是当前流行的 Java Web 前后端开发利器。本书以相应软件的新版本为平台，基于贯穿式编写模式，以这三部分内容为侧重点，注重前后衔接，精心选择了基础知识点、核心知识点和扩展知识点进行介绍。全书在知识点的讲解＋场景应用与挑战中贯穿知识的融合，渐进式地引领读者深刻理解、掌握和使用，为从事实际应用开发建立清晰的技术思路、扎实的知识技能。

本书由浅入深，通俗易懂，注重理论联系实际。本书适用于没有 Java Web 编程经验的初学者，也适合具有一定编程基础、需要提高实践能力的开发人员作为参考用书，还适合作为各类学校相关课程的教材。

本书封面贴有清华大学出版社防伪标签，无标签者不得销售。
版权所有，侵权必究。举报：010-62782989，beiqinquan@tup.tsinghua.edu.cn。

图书在版编目(CIP)数据

jQuery+Vue.js+Spring Boot 贯穿式项目实战：微课视频版/陈冈编著. —北京：清华大学出版社，2022.3
(21 世纪项目案例开发丛书)
ISBN 978-7-302-60117-3

Ⅰ. ①j… Ⅱ. ①陈… Ⅲ. ①JAVA 语言－程序设计 ②网页制作工具－程序设计 Ⅳ. ①TP312.8
②TP393.092.2

中国版本图书馆 CIP 数据核字(2022)第 021027 号

责任编辑：闫红梅
封面设计：刘　键
责任校对：胡伟民
责任印制：杨　艳

出版发行：清华大学出版社
　　网　　址：http://www.tup.com.cn, http://www.wqbook.com
　　地　　址：北京清华大学学研大厦 A 座　　邮　编：100084
　　社 总 机：010-83470000　　邮　购：010-83470235
　　投稿与读者服务：010-62776969, c-service@tup.tsinghua.edu.cn
　　质量反馈：010-62772015, zhiliang@tup.tsinghua.edu.cn
　　课件下载：http://www.tup.com.cn, 010-83470236
印　刷　者：北京富博印刷有限公司
装　订　者：北京市密云县京文制本装订厂
经　　销：全国新华书店
开　　本：203mm×260mm　　印　张：19　　字　数：472 千字
版　　次：2022 年 4 月第 1 版　　印　次：2022 年 4 月第 1 次印刷
印　　数：1～2000
定　　价：69.80 元

产品编号：094265-01

前言
FOREWORD

目前,jQuery 在市场上有非常广泛的应用,Vue.js 成为三大著名前端开发框架之一,Spring Boot 则是 Java EE 开发的重器。三者结合,在当今 Java Web 开发领域,具有很好的技术发展前景。

在长期软件开发和教学过程中,作者深切感受到找到一本合适的 Java Web 参考书并不容易。目前,市场上这类书很多:有的讲得非常细致,像是相关软件的指南、手册。但过于细致,对学习者(特别是自学者)并不是好事,更不适合作为教材使用;有的非常着重于技术性开发,内容烦琐,大而全,并不注重知识的融合、贯穿,且过于渗透技术,阅读起来很不轻松。就像有些学员在学车的过程中,教练常常讲开车就是要讲求"人车合一",有些参考书亦如此。

有鉴于此,作者在编写本书的过程中始终把握以下这三点:

❖ 夯实基础,精心取舍内容,轻松进阶。没有一本书能够解决所有知识点。当然,基础知识是必须介绍的,但是不追求大而全的繁杂,也不追求过于细致的讲解。舍得,有舍才会有得。必须精心选择,有所取舍,例如介绍 JSP 内容时,JavaBean 的内容就舍弃了,但 EL 表达式没有舍弃,目的是通过其建立初步的分层思想,为后续 MVC 打下基础。另外,特别注重内容的易理解性。除了大量注释,还精选示例,例如第 10 章场景应用中的动态增删书目,就能让读者充分体会 Vue 是如何通过数据的改变来驱动页面的变化。

❖ 多技术融合,贯穿使用,兼顾前沿。谁都知道,学以致用。而实践中问题的解决,往往靠的并不是某一种知识技能。学了要更好地用,就必须强调知识的融合。很多人发现,本书目录与同类书有较大差异,作者精心编排了章节顺序:学了 MVC 后,很多人其实还是比较模糊的,难以用起来。那么,接着学习 jQuery,用其实现 V 层的交互处理,有了初步感觉。紧接着,再学习 Spring Boot,用其实现 C 层、M 层的处理。然后,才安排学习 Vue。用 Vue 来重现前面案例的 M 层内容。这样处理,不但能更好地理解 MVC 设计模式,又会对 jQuery、Vue 的差异性有不错的体会,对 Spring Boot 所扮演的角色有清晰的认识。再例如,第 8 章场景应用中的学生信息查询,着重用 jQuery 实现前端的视图层处理,而第 9 章用 Spring Boot 实现学生信息查询,侧重的是后端实现。这样处理,读者就会对前后端有更清晰的理解。而兼顾前沿这点,也贯穿其中:书中专门介绍了分布式缓存网格、R2DBC、响应式处理、函数式编程等内容。具体来看,例如第 12 章基于流行方式实现 WebSocket 聊天处理,第 13 章则改用 Spring WebFlux 实现;第 5 章文件上传采用常规方式实现,而第 13 章则用响应式数据流技术来实现。

❖ 渐进式推进,场景应用,项目实战。章节安排,由浅入深,逐步向不太容易理解的知识推进。每章安排了融合本章知识的场景应用,以及精心选择的场景挑战。有些章节的场景挑战,其解决方法，其实暗含在后面某些示例或者场景任务中。最后一章，提供了一个完整的综合性项目案例，对前面所学知识融合运用，但又有推进与提升。

本书共分 13 章，主要内容包括：第 1 章讲解如何构建好开发环境；第 2 章介绍 HTML 基础知识；第 3 章介绍 CSS；第 4 章以 JavaScript 为主要内容；第 5 章讲解 JSP 基础知识；第 6 章熟悉 MVC 设计模式；第 7 章探究数据库连接池；第 8 章介绍 jQuery；第 9 章重点出击 Spring Boot；第 10 章侧重于 Vue 3.x；第 11 章用图形实现数据可视化；第 12 章介绍消息服务；第 13 章提供一个基于 Spring WebFlux 的项目开发案例。

在本书的编写过程中得到了清华大学出版社的大力支持，在此深表谢意！感谢家人提供的暖心支持，让我得以顺利完成本书的编写！

本书可满足各类学校相关课程的教学需要，也可作为广大技术开发者的参考用书。在编写过程中，作者虽力求精益求精，但也难免存在一些疏漏或不足之处，敬请读者批评指正！

感谢您使用本书，希望本书能够成为您的良师益友！

编　者

2021 年 9 月

目录
CONTENTS

第 1 章　Java Web 开发基础 ······················· 1

　1.1　Java Web 概述 ································· 1
　1.2　C/S 与 B/S 模式 ······························· 1
　1.3　搭建开发环境 ································· 2
　　1.3.1　安装 Java SE JDK ······················· 3
　　1.3.2　安装 Apache Tomcat ···················· 4
　　1.3.3　安装 PostgreSQL 数据库 ················ 7
　　1.3.4　使用 IntelliJ IDEA ······················ 11
　1.4　创建 Maven Web 站点 ······················· 15
　　1.4.1　Maven 简介 ···························· 15
　　1.4.2　修改资源仓库的镜像地址 ·············· 15
　　1.4.3　创建 Maven WebApp 站点项目 ········ 16
　　1.4.4　修改项目编译版本 ····················· 18
　1.5　场景任务挑战——配置自己的站点 ········· 19

第 2 章　HTML 基础 ······························· 20

　2.1　HTML 简介 ···································· 20
　2.2　HTML 文档基本结构 ························· 20
　2.3　头部和主体标签 ······························ 21
　　2.3.1　头部标签＜head＞ ····················· 21
　　2.3.2　主体标签＜body＞ ····················· 21
　2.4　其他常用标签 ································· 22
　　2.4.1　链接分段与换行 ························ 22
　　2.4.2　表格和列表 ····························· 23
　　2.4.3　层标签和组合标签 ····················· 25
　　2.4.4　图像和媒体 ····························· 26
　　2.4.5　对话框 ·································· 27
　　2.4.6　表单及表单元素 ························ 28

2.4.7　模板 …… 31
　　　2.4.8　内联框架 …… 31
　2.5　事件 …… 31
　2.6　场景应用示例——显示字符映射表图标 …… 32
　　　2.6.1　应用需求 …… 32
　　　2.6.2　实现思路 …… 33
　　　2.6.3　具体实现 …… 34
　2.7　场景任务挑战——注册与叠加的层 …… 34

第3章　CSS（层叠样式表） …… 35

　3.1　CSS简介 …… 35
　3.2　CSS基础 …… 35
　　　3.2.1　CSS基本格式 …… 35
　　　3.2.2　应用方式 …… 36
　3.3　CSS样式设置 …… 38
　　　3.3.1　文本 …… 38
　　　3.3.2　背景 …… 39
　　　3.3.3　边框和边距 …… 40
　　　3.3.4　定位溢出和浮动 …… 40
　　　3.3.5　伪类和伪元素 …… 41
　　　3.3.6　多列 …… 43
　　　3.3.7　动画 …… 45
　3.4　场景应用示例——功能导航条 …… 46
　　　3.4.1　应用需求 …… 46
　　　3.4.2　实现思路 …… 47
　　　3.4.3　CSS代码 …… 47
　3.5　场景任务挑战——导航菜单 …… 48

第4章　JavaScript脚本语言 …… 49

　4.1　JavaScript简介 …… 49
　4.2　JavaScript的使用 …… 49
　　　4.2.1　页面直接使用 …… 49
　　　4.2.2　使用脚本文件 …… 50
　4.3　变量和常量 …… 50
　　　4.3.1　使用var和let声明变量 …… 50
　　　4.3.2　使用const声明常量 …… 51
　4.4　基本数据类型 …… 51
　4.5　函数 …… 52

4.5.1 使用 function 定义函数 ·················· 52
4.5.2 使用箭头函数 ·················· 52
4.6 数组 ·················· 53
4.7 对象 ·················· 54
4.7.1 对象概述 ·················· 54
4.7.2 当前对象 this ·················· 55
4.7.3 窗口对象 Window ·················· 55
4.7.4 文档对象 Document ·················· 57
4.7.5 事件状态 Event ·················· 59
4.7.6 页面定位 Location ·················· 59
4.7.7 样式处理 Style ·················· 60
4.7.8 对象包装器 Object ·················· 60
4.8 异步操作 Promise ·················· 61
4.8.1 Promise 对象 ·················· 61
4.8.2 async 和 await ·················· 62
4.9 控制语句 ·················· 63
4.9.1 导入(import)和导出(export) ·················· 63
4.9.2 条件判断 if…else ·················· 64
4.9.3 多条件分支 switch ·················· 65
4.9.4 循环操作 for ·················· 66
4.9.5 do…while 和 while 语句 ·················· 69
4.9.6 try…catch…finally 语句 ·················· 69
4.10 表单数据 FormData ·················· 70
4.10.1 通过表单 <form> 创建 ·················· 70
4.10.2 用代码生成 FormData ·················· 71
4.10.3 处理文件数据 ·················· 71
4.11 使用 JSON ·················· 73
4.11.1 JSON 简介 ·················· 73
4.11.2 JSON 基本语法 ·················· 73
4.11.3 解析为 JSON 对象 ·················· 75
4.11.4 转换为 JSON 字符串 ·················· 75
4.12 场景应用示例——动态增删书目 ·················· 76
4.12.1 应用需求 ·················· 76
4.12.2 处理思路 ·················· 77
4.12.3 实现 HTML 页面 ·················· 77
4.12.4 编写 JS 脚本文件 ·················· 78
4.13 场景任务挑战——勾选删除 ·················· 78

第 5 章 JSP 基础 79

5.1 JSP 概述 79
5.1.1 JSP 简介 79
5.1.2 JSP 基本页面结构 79
5.1.3 配置 Tomcat 依赖 80

5.2 JSP 基本语法 80
5.2.1 程序段 80
5.2.2 表达式 81
5.2.3 JSP 中的注释 81

5.3 JSP 内置对象 82
5.3.1 out 82
5.3.2 request 83
5.3.3 response 84
5.3.4 session 85
5.3.5 application 87

5.4 使用 Servlet 88
5.4.1 Servlet 简介 88
5.4.2 Servlet 生命周期 88
5.4.3 doGet()和 doPost()方法 89
5.4.4 加入 Servlet 依赖 89
5.4.5 创建 Servlet 90

5.5 EL 表达式语言 91
5.5.1 EL 概述 91
5.5.2 加入 JSTL 依赖 92
5.5.3 内置对象 92
5.5.4 条件输出 92
5.5.5 循环输出 94

5.6 监听器 95
5.6.1 监听器类型 95
5.6.2 基于监听器的在线用户统计 95

5.7 与数据库交互 96
5.7.1 创建 users 表并加入数据库依赖 96
5.7.2 数据库连接 97
5.7.3 JDBC 应用 98

5.8 场景应用示例 102
5.8.1 文件上传 102
5.8.2 在页面中显示 Excel 表格 103

5.8.3 用 PDF 显示古诗 ……………………………………………………………… 106
5.9 场景任务挑战——有背景图的 PDF 古诗 ………………………………………… 108

第 6 章　MVC 设计模式 ……………………………………………………………… 109

6.1 MVC 概述 ……………………………………………………………………………… 109
  6.1.1 传统 JSP 开发模式 ……………………………………………………………… 109
  6.1.2 MVC 原理 ………………………………………………………………………… 109
  6.1.3 MVC 的优缺点 …………………………………………………………………… 110
6.2 MVC 实现过程 ………………………………………………………………………… 111
6.3 场景应用示例——用户注册 ………………………………………………………… 111
  6.3.1 应用需求 ………………………………………………………………………… 111
  6.3.2 处理思路 ………………………………………………………………………… 112
  6.3.3 模型层 …………………………………………………………………………… 112
  6.3.4 控制器层 ………………………………………………………………………… 115
  6.3.5 视图层 …………………………………………………………………………… 115
6.4 场景任务挑战——学生信息查询 …………………………………………………… 117

第 7 章　数据库连接池 ………………………………………………………………… 118

7.1 连接池概述 …………………………………………………………………………… 118
  7.1.1 连接池基本原理 ………………………………………………………………… 118
  7.1.2 常见连接池产品 ………………………………………………………………… 119
  7.1.3 Tomcat 连接池示例 ……………………………………………………………… 119
7.2 HikariCP 连接池 ……………………………………………………………………… 121
  7.2.1 HikariCP 简介 …………………………………………………………………… 121
  7.2.2 加入 HikariCP 依赖 ……………………………………………………………… 121
  7.2.3 配置 HikariCP 连接池 …………………………………………………………… 121
  7.2.4 查看 HikariCP 活动情况 ………………………………………………………… 122
7.3 场景应用示例——优化 HikariCP 使用 …………………………………………… 123
  7.3.1 应用需求 ………………………………………………………………………… 123
  7.3.2 创建监听器类 AppService ……………………………………………………… 124
  7.3.3 连接池的构建和关闭 …………………………………………………………… 124
  7.3.4 修改 DBFactory 类 ……………………………………………………………… 124
7.4 场景任务挑战——动态配置 HikariCP ……………………………………………… 125

第 8 章　jQuery 前端开发 …………………………………………………………… 126

8.1 jQuery 概述 …………………………………………………………………………… 126
  8.1.1 jQuery 简介 ……………………………………………………………………… 126
  8.1.2 jQuery 的使用 …………………………………………………………………… 126

8.1.3　jQuery 基础语法 ……………………………………………………………… 128
8.2　jQuery 选择器 …………………………………………………………………………… 128
　　8.2.1　元素选择器 …………………………………………………………………… 128
　　8.2.2　属性选择器 …………………………………………………………………… 130
　　8.2.3　CSS 选择器 …………………………………………………………………… 131
8.3　jQuery 操作 ……………………………………………………………………………… 132
　　8.3.1　元素操作 ……………………………………………………………………… 132
　　8.3.2　属性操作 ……………………………………………………………………… 133
　　8.3.3　操作 CSS 类 …………………………………………………………………… 134
　　8.3.4　遍历操作 ……………………………………………………………………… 135
　　8.3.5　事件函数 ……………………………………………………………………… 136
8.4　jQuery 动画 ……………………………………………………………………………… 137
8.5　与服务器交互 …………………………………………………………………………… 137
　　8.5.1　用 Jackson 格式化数据 ……………………………………………………… 138
　　8.5.2　$.ajax ………………………………………………………………………… 138
　　8.5.3　$.get 和 $.getJSON …………………………………………………………… 140
　　8.5.4　$.post ………………………………………………………………………… 141
8.6　场景应用示例 …………………………………………………………………………… 143
　　8.6.1　下拉选择框联动 ……………………………………………………………… 143
　　8.6.2　学生信息查询 ………………………………………………………………… 144
8.7　场景任务挑战——动态增删书目 ……………………………………………………… 148

第 9 章　Spring Boot 开发基础 ………………………………………………………… 149

9.1　RESTful 概述 …………………………………………………………………………… 149
　　9.1.1　REST 简介 …………………………………………………………………… 149
　　9.1.2　RESTful 要义 ………………………………………………………………… 149
　　9.1.3　RESTful 请求风格 …………………………………………………………… 150
9.2　Spring Boot 概述 ………………………………………………………………………… 150
　　9.2.1　Spring Boot 简介 ……………………………………………………………… 150
　　9.2.2　创建 Spring Web MVC 项目 ………………………………………………… 150
　　9.2.3　Spring Boot 入口类 …………………………………………………………… 153
　　9.2.4　配置 HikariCP 连接池 ………………………………………………………… 153
　　9.2.5　Spring Boot 常用注解 ………………………………………………………… 154
　　9.2.6　JpaRepository 数据访问 ……………………………………………………… 155
9.3　Reactive 响应式处理 …………………………………………………………………… 162
　　9.3.1　响应式概述 …………………………………………………………………… 162
　　9.3.2　Reactor 基本原理 ……………………………………………………………… 162
　　9.3.3　Reactor 核心包 publisher …………………………………………………… 162

9.3.4 单量 Mono＜T＞ ... 163
9.3.5 通量 Flux＜T＞ ... 163
9.3.6 并行 ParallelFlux ... 165
9.3.7 处理槽 Sinks ... 166
9.3.8 响应式 R2dbcRepository ... 166
9.3.9 启用响应式 R2DBC ... 167
9.3.10 启用分布式内存网格 ... 168
9.4 Spring WebFlux ... 170
9.4.1 Spring WebFlux 简介 ... 170
9.4.2 WebFlux 应用的入口类 ... 171
9.4.3 配置 WebFlux 应用 ... 171
9.4.4 HandlerFilterFunction 操作过滤 ... 172
9.4.5 HandlerFunction 业务处理 ... 172
9.4.6 RouterFunction 路由函数 ... 173
9.4.7 Multipart Data 多域数据 ... 173
9.5 场景应用示例 ... 174
9.5.1 学生信息查询 ... 174
9.5.2 基于 JWT 令牌实现分布式登录 ... 177
9.6 场景任务挑战——模糊查询 ... 184

第 10 章　Vue.js 渐进式框架 ... 185

10.1 Vue 概述 ... 185
10.2 Vue 应用基础 ... 186
10.2.1 创建 Vue 应用 ... 186
10.2.2 生命周期 ... 187
10.2.3 组合式函数 setup() ... 188
10.2.4 响应性函数 ... 189
10.2.5 解构 ... 189
10.3 基础语法 ... 190
10.3.1 模板语法 ... 190
10.3.2 计算属性和侦听 ... 193
10.3.3 表单域的数据绑定 ... 196
10.3.4 组件对象的数据绑定 ... 197
10.3.5 事件绑定和触发 ... 198
10.3.6 自定义元素 defineCustomElement ... 200
10.3.7 条件和列表渲染 ... 201
10.4 h() 函数和渲染函数 render() ... 203
10.5 使用组件 ... 205

10.5.1 组件定义及动态化 205
10.5.2 异步组件 206
10.6 函数集 207
10.7 使用 Axios 请求后端数据 207
10.7.1 Axios 简介 207
10.7.2 请求响应结构和错误处理 208
10.7.3 发起请求 209
10.7.4 配置拦截器 209
10.8 场景应用示例 209
10.8.1 动态增删书目 209
10.8.2 学生信息查询 211
10.9 场景任务挑战——下拉选择框联动 213

第 11 章 用图形展示数据 214

11.1 Web 数据的图形可视化 214
11.2 Apache ECharts 图形前端 214
11.2.1 Apache ECharts 简介 214
11.2.2 下载与引用 215
11.2.3 ECharts 创建图形的架构 215
11.3 JFreeChart 图形后端 217
11.3.1 JFreeChart 简介 217
11.3.2 加入 JFreeChart 相关依赖 217
11.3.3 JFreeChart 应用基础 217
11.4 场景应用示例 219
11.4.1 招生情况 SVG 饼图（JFreeChart） 220
11.4.2 招生情况面积图（ECharts） 224
11.5 场景任务挑战——招生情况直方图 228

第 12 章 消息服务 229

12.1 消息服务概述 229
12.1.1 消息服务简介 229
12.1.2 消息服务模式 229
12.2 用 Apache Kafka 作为消息服务器 231
12.2.1 Apache Kafka 简介 231
12.2.2 启用 Kafka 服务器 231
12.2.3 Kafka 配置和管理 232
12.2.4 KafkaTemplate 模板 233
12.2.5 生产者 Producer 和消费者 Consumer 233

12.2.6　Kafka 响应式发送器和接收器 ………………………………… 234
　12.3　整合 WebSocket 及 SockJS ………………………………………… 235
　　　12.3.1　在客户端使用 ………………………………………………… 235
　　　12.3.2　在服务端使用 ………………………………………………… 236
　　　12.3.3　使用拦截器 …………………………………………………… 237
　　　12.3.4　Spring WebFlux 中的 WebSocket …………………………… 237
　12.4　场景应用示例——聊天室 …………………………………………… 238
　　　12.4.1　应用需求 ……………………………………………………… 238
　　　12.4.2　主页 …………………………………………………………… 239
　　　12.4.3　登录组件 users.component.js ……………………………… 240
　　　12.4.4　登录后端处理 ………………………………………………… 241
　　　12.4.5　聊天组件 chat.component.js ………………………………… 242
　　　12.4.6　实现 JWT 令牌验证 ………………………………………… 244
　　　12.4.7　配置 Kafka 和 WebSocket 全局参数 ……………………… 245
　　　12.4.8　WebSocket 配置类及拦截器 ………………………………… 245
　　　12.4.9　创建聊天服务 ………………………………………………… 246
　12.5　场景任务挑战——学生、教师各自的聊天室 ……………………… 247

第 13 章　教务辅助管理项目开发 ……………………………………… 248

　13.1　系统概述 ……………………………………………………………… 248
　13.2　系统功能简介 ………………………………………………………… 248
　13.3　系统技术选型 ………………………………………………………… 249
　　　13.3.1　前端组件化 …………………………………………………… 249
　　　13.3.2　后端模块化 …………………………………………………… 250
　13.4　数据表设计 …………………………………………………………… 250
　13.5　系统实现 ……………………………………………………………… 251
　　　13.5.1　创建 Spring Reactive Web 项目 ……………………………… 251
　　　13.5.2　配置 application.yml 全局参数 ……………………………… 251
　　　13.5.3　加入项目主要依赖 …………………………………………… 252
　　　13.5.4　引入 JS 支持文件 …………………………………………… 253
　　　13.5.5　使用聚合器管理组件 ………………………………………… 253
　　　13.5.6　应用入口程序 ………………………………………………… 254
　　　13.5.7　WebFlux 配置和路由配置 …………………………………… 255
　　　13.5.8　身份验证过滤组件 …………………………………………… 257
　　　13.5.9　主页 …………………………………………………………… 258
　　　13.5.10　用户登录 ……………………………………………………… 259
　　　13.5.11　消息推送 ……………………………………………………… 262
　　　13.5.12　用户注册 ……………………………………………………… 265

13.5.13　学院风采 …………………………………………………………… 268
　　13.5.14　学生信息模糊查询 …………………………………………………… 271
　　13.5.15　招生数据一览 ………………………………………………………… 276
　　13.5.16　资料上传 ……………………………………………………………… 280
　　13.5.17　交流空间 ……………………………………………………………… 283
13.6　打包发布 …………………………………………………………………………… 285

第 1 章

Java Web开发基础

本章主要介绍 Java Web 基本开发模式、开发环境配置等内容，并创建一个基于 Maven 的 Java Web 站点，为后续章节的学习打下坚实的基础。

1.1 Java Web 概述

Java 是一种编程语言，Web 则是"网络"之意。

在网络编程领域，Java 是卓越的开发语言之一。在著名的 TIOBE 网站发布的编程语言排行榜中，Java 长期处于前三的位置。Java 以其严谨、应用生态好、社区强大著称，具有巨大的影响力。长期以来，Java 为 Web 编程领域的发展注入了强大的推动力。Java Web 一直在快速适应编程环境及程序员编程方法的变化，因此在 Web 开发领域，用 Java 技术解决现实中的问题，具有非常广泛的应用基础。

1.2 C/S 与 B/S 模式

在现实中使用的各种软件，绝大部分都需要网络环境的支撑，才能正常运行，例如安卓手机上的很多 App、各种类型的网站，或者企事业单位内部的应用系统等。这些应用中的某些软件，就是用 Java 编写的，那么它们在运行模式上有什么差异呢？这就涉及 C/S 模式和 B/S 模式。

1. C/S（Client/Server）模式

C/S 模式（也称为 C/S 结构）是由美国 Borland 公司最早研发的，它以网络环境为基础，将计算机应用分散到多台计算机以提高系统性能。在 C/S 结构中，所有的数据都存放在服务器上（因此也称为数据库服务器），客户机只负责向服务器提出用户的处理需求，由服务器完成对数据的处理后将结果通过网络传递给客户端计算机。C/S 模式如图 1-1 所示。

整个过程，可以打比方描述为：你（客户端）到银行柜台申请办理新银行卡业务（发出请求），柜台工作人员（服务器）响应了你的要求，从存放银行卡的卡柜（数据库服务器）取出一张银行卡，为你办理完成了新银行卡业务。

C/S 模式的优势在于:可以充分利用客户端设备本身的处理能力,减轻了应用服务器的处理压力。其劣势在于:软件通常需要先下载安装到客户端设备。当软件需要更新换代时,往往需要对所有的客户端及应用服务器进行同步升级。因此,C/S 系统的维护成本相对较高,技术较为复杂。

现实中有很多应用软件,例如 QQ、微信、抖音、喜马拉雅、超市收银系统等,就是采用 C/S 模式。

2. B/S(Browser/Server)模式

B/S 是对 C/S 结构的一种提升,是随着 Internet 技术的发展而逐步流行起来的一种网络应用模式。在这种结构的系统中,包含两大部分:Web 客户端、Web 服务器端。Web 客户端一般通过浏览器向用户提供各种操作界面。这些操作界面,主要用于数据处理结果的展示,或者前期预处理操作(例如验证用户输入的 E-mail 地址格式是否正确);Web 服务器端则接受用户对数据的操作请求并对请求进行各种相应处理,再根据需要传递给数据库服务器。B/S 模式的结构如图 1-2 所示。

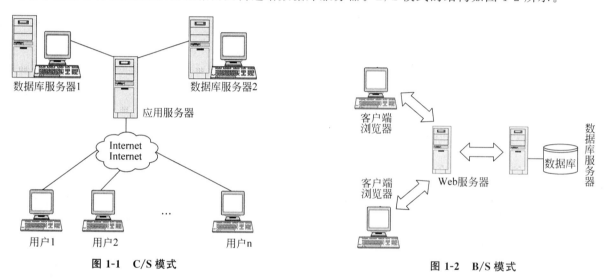

图 1-1　C/S 模式　　　　　　　　　　图 1-2　B/S 模式

B/S 模式的优势在于:客户端要求简单,只需要浏览器即可,这使得在网络上扩充系统功能变得非常容易,通常只需要更改服务器端即可,客户端并不需要升级处理。但是,当客户端用户比较多时,服务器运行负荷会比较重,可能会导致服务器反应缓慢甚至"崩溃"等问题,这是 B/S 模式最大的劣势。

可以将基于 B/S 模式的应用系统简单归纳为两类:一类是现实中的各类网站,例如京东、天猫、12306 等。另一类是企事业单位使用的各种应用系统,例如招考办的考生志愿填报系统,公司设备管理网等;这一类与我们习惯中所称的"网站"有差别,它们实际上是基于浏览器运行的用于信息管理目的的 Web 应用系统。

综上所述,Java Web 一般是指基于 B/S 模式,用 Java 技术来解决相关 Web 互联网领域应用问题的技术总和。

1.3　搭建开发环境

"工欲善其事,必先利其器",这部分内容比较重要,是我们进行 Java Web 开发的起点。

1.3.1 安装 Java SE JDK

1. 下载

Java SE JDK 是 Java Platform Standard Edition Development Kit 的缩写（常常简称为 JDK），是构建各类 Java 应用的基础环境。JDK 版本更新比较快，使用市场上的主流版本（JDK8）而非最新版，是一种比较好的选择，目前 JDK8 的最新版本是 JDK-8u291。这里的 u 是 update 之意，291 是指 JDK8 当前更新编号，下次更新也许会变成 JDK-8u292。

可以从 Qracle 官网下载 JDK8，如图 1-3 所示。

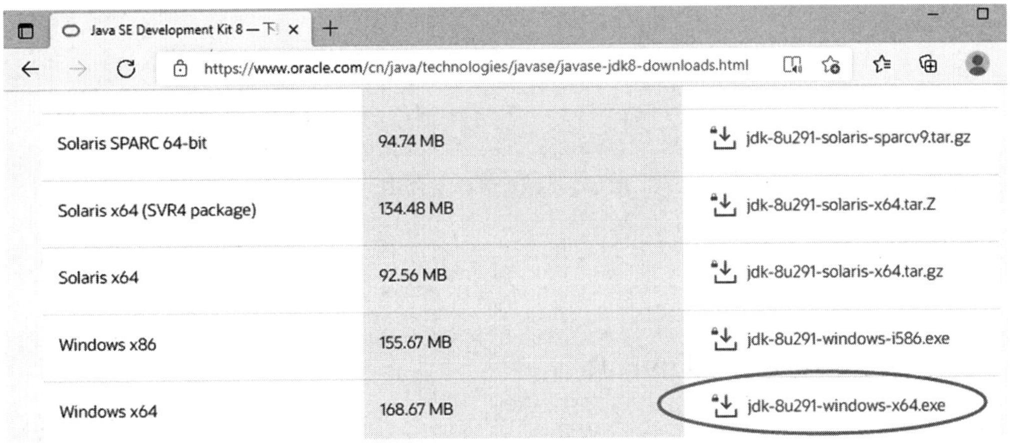

图 1-3　JDK 下载

按照页面提示，注册后就可以下载。我们下载的是 JDK8 的 64 位版本。

2. 安装

单击运行下载得到的 jdk-8u291-windows-x64.exe，按照提示进行安装。有两个地方需要注意：
（1）为了后续方便起见，我们将 JDK8 安装到 D:\Java\jdk8，如图 1-4 所示。

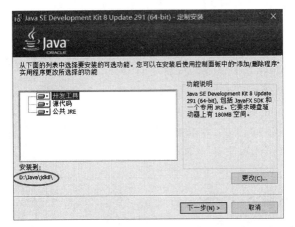

图 1-4　JDK 目标文件夹

(2) 接下来,选择 JRE 安装的目标文件夹,如图 1-5 所示。这里,我们将其安装到 D:\Java\jre8,单击"下一步"按钮,安装很快即可完成。注意:需要事先在 D:\Java 下创建好 jre8 文件夹。

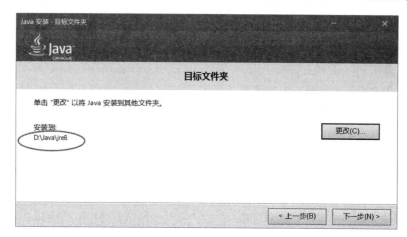

图 1-5　JRE 目标文件夹

有 Java 知识基础的读者,会在很多教程里经常看到如何设置 Java 环境变量的介绍,这里却并没有这样处理。提示:Java 环境变量的设置,并不是必需的!

1.3.2　安装 Apache Tomcat

从前面 B/S 模式的叙述中得知,Java Web 系统的运行需要 Web 服务器的支持。市场上的 Web 服务器很多,例如 Resin、Tomcat、Weblogic、Undertow、Jetty 等。在此,我们选用 Apache Tomcat 作为 Web 服务器。

1. Tomcat 简介

Tomcat 是一款开源、免费的 Web 服务器产品,源自著名的 Apache 软件基金会。根据埃文斯数据公司(Evans Data Corporation) 的数据显示,目前已有超过 70% 的企业采用了 Tomcat。

可以到 Apache Tomcat 的官网下载 Tomcat。

2. Tomcat 的下载和安装

下载 Tomcat9 作为 Web 服务器。打开 Tomcat 官网,单击界面左边的 Download→Tomcat 9,然后选择下载 64 位压缩版本,如图 1-6 所示。

打开下载得到的 apache-tomcat-9.0.50-windows-x64.zip,解压到某个磁盘,例如 D 盘,安装完成。这里,"D:\apache-tomcat-9.0.50"文件夹是本书使用的 Tomcat 安装文件夹。

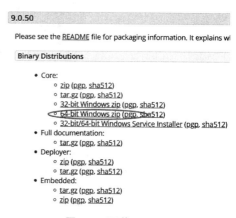

图 1-6　下载 Tomcat

3. 安装为 Windows 服务(可选)

有时候为了简便起见,可以将 Tomcat 安装为 Windows 的服务项目。注意:安装为服务并不是必需的，可以略过这一步骤。这样做的目的，是为了给以后的某些处理带来方便。安装和卸载 Tomcat 服务的命令格式如下：

安装服务：service install Tomcat9

卸载服务：service uninstall Tomcat9

上面命令中的"Tomcat9"是服务名称，可以自行指定。现在，将第 2 步中的 Tomcat 安装为 Windows 服务。首先，打开 Windows 命令提示符，切换到文件夹 D:\apache-tomcat-9.0.50\bin，输入命令：service install Tomcat9，结果如图 1-7 所示。

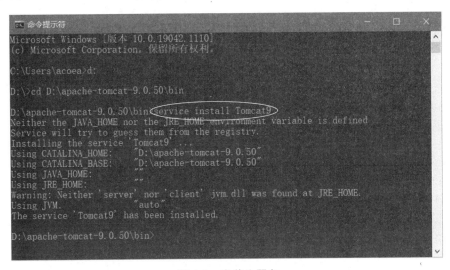

图 1-7　安装为服务

按回车键，会提示服务已安装。通过"控制面板"→"系统和安全"→"管理工具"→"服务"，打开 Windows 的服务，就可以很方便地启动/停止 Tomcat 服务，如图 1-8 所示。

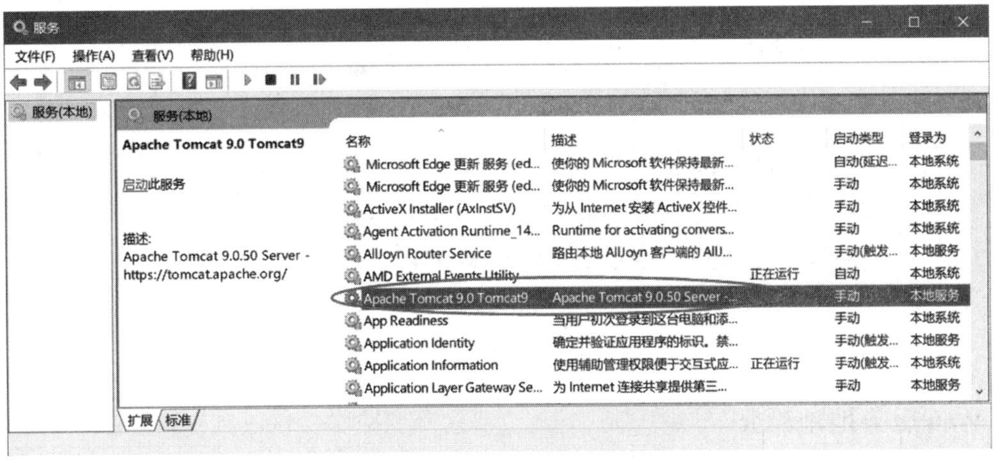

图 1-8　Tomcat 服务

提示：如果 Tomcat 服务无法启动，可以在服务上右击鼠标，在弹出的快捷菜单中选择"属性"→"登录"。单击选中"本地系统账户"单选按钮，即将原来默认选择的"此账户"修改为"本地系统账户",单击"确定"按钮,如图 1-9 所示,再重启 Tomcat 服务即可。当然,能够正常启动的,无须做此修改。

4. 修改日志编码

用记事本打开 D:\apache-tomcat-9.0.50\conf\logging.properties 文件,将下面这句代码中的 UTF-8,修改为 GB18030(国家汉字编码字符集),然后保存。

```
java.util.logging.ConsoleHandler.encoding = UTF-8
```

这句代码设置了 Tomcat 运行时日志输出使用的编码,改为 GB18030 后可以避免日志输出中的汉字乱码问题。

5. 测试 Tomcat 服务器

先启动 Tomcat 服务,再在浏览器中输入:http://localhost:8080,将打开 Tomcat 服务器默认的主页,出现如图 1-10 所示的页面,说明 Tomcat 已经成功运行。

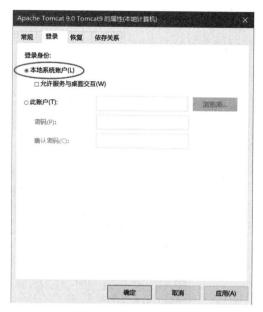

图 1-9 修改登录方式

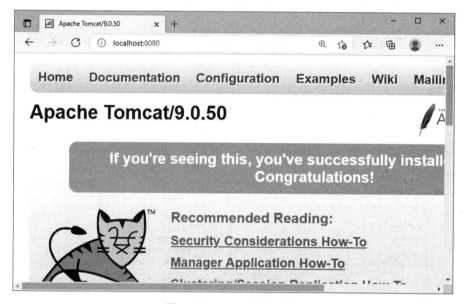

图 1-10 Tomcat 服务主页

6. Web 服务地址简介

通过在浏览器地址栏输入 http://localhost:8080/,可以成功打开本机 Tomcat 服务器的主页。那么,地址格式是怎样定义的?

格式:协议://服务器名:端口号/路径

下面以 http://localhost:8080 为例来详细解读。

1）协议

这里使用的是 http 协议。此外，还有 https、tel、ftp、ws 等多种协议。协议是计算机之间进行联络的一种约定，打个比方：上课时老师与学生之间约定的语言是普通话（协议）而非粤语（另外一种协议）。

2）服务器名

服务器名是 localhost，也可以用 IP 地址。由于 Tomcat 目前运行在本机上，也就是说我们访问的是本机的 Web 服务器，所以也可以用本机的 IP 地址"127.0.0.1"代替 localhost，即用 http://127.0.0.1:8080 打开 Tomcat 服务主页。如果要访问别人计算机上的 Tomcat 主页，就不能用 localhost 了，需要用别人计算机的 IP 地址替换 localhost 才行。

3）端口号

8080，这是 Tomcat 默认的端口号。端口号表示计算机与外部进行通信的出口，外部则通过这个出口与计算机进行通信。因此，要与 Tomcat 提供的 Web 服务进行联系，需要通过 8080 端口才行。打个比方：我们在 7 号教学楼 108 教室上"Java Web 程序设计"这门课，在 109 教室上"数据库原理"这门课。7 号楼就类似于 Tomcat 服务器，108 教室类似于端口号 8080，而 109 教室则代表 Tomcat 提供的另外一种服务出口。如此看来，服务器名加上端口号一起，就能够表示对应网址是由哪台服务器提供的哪种服务。

端口号也可以修改，打开 D:\apache-tomcat-9.0.50\conf\server.xml 文件，找到：

```
<Connector port="8080" protocol="HTTP/1.1"
           connectionTimeout="20000"
           redirectPort="8443" />
```

可将 8080 修改为其他端口号，例如 8081，修改后重启 Tomcat 服务即可生效。

看到这里，有读者可能会有疑问：我们平时输入各种网址，并不需要输入端口号呀？这是因为，这些网址使用的是 HTTP 默认的端口号 80，而 80 是不需要显式地输入的。

4）路径

"/"代表了 Tomcat 主页的根目录。

现在，相信对于类似 http://www.myweb.edu.cn/em/01/0704 这样的网址，大家应该很清楚其中的含义了。

1.3.3 安装 PostgreSQL 数据库

现实中的大部分网站，都提供了数据库的支持。有了数据库支持，各种 Web 系统就能够更好地提供各种 Web 数据服务。

1. PostgreSQL 数据库简介

很多读者听说过 Access、SQL Server、Oracle、MySQL 数据库，但对 PostgreSQL 则有些陌生，这并不妨碍 PostgreSQL 成为一款伟大的产品。PostgreSQL 数据库源自美国加州大学伯克利分校计算机系开发的 Postgres。PostgreSQL 的标签就是：强大、稳定、最高级的开源 DBMS。

提示：PostgreSQL 简称 PG。官方拼读为"post-gress-Q-L"，经常被简略念为"postgres"。

2. 为什么选择 PostgreSQL

PG 是世界领先的企业级数据库产品，开源、免费。与昂贵的 Oracle 数据库相比，它可以成为很好的替代方案。

PG 已经成为大数据、云计算领域中关系数据库存储管理的最佳选择之一。PG 在企业应用中非常广泛，其客户涵盖包括世界五百强在内的众多国内外大型企业；PG 的行业应用也非常广泛，涵盖金融、能源、零售、IT、互联网等众多行业。

PG 在全球得到很多著名企业的青睐。国内客户，例如：百度、阿里巴巴、腾讯、去哪儿网、高德地图、中国邮政、顺丰速递、华为、京东、中国移动、华夏银行、光大银行等。国外客户，例如：迈克菲杀毒软件、东芝、索尼、亚马逊、日本电报电话公司等。

3. 下载和安装

1）下载

可以去 PG 官网下载，我们下载 13.3 这个版本，这个只有 64 位版。如图 1-11 所示。

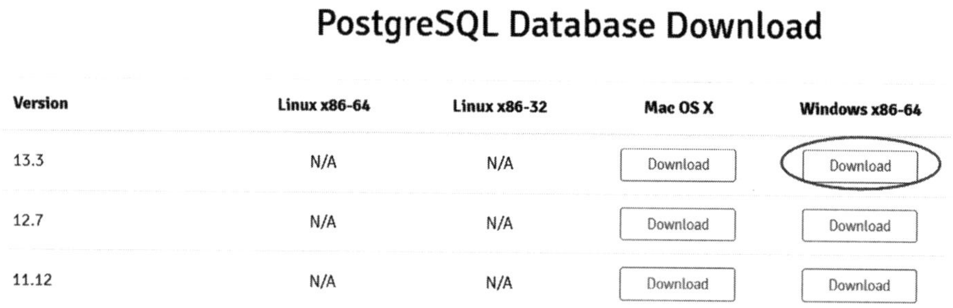

图 1-11 PostgreSQL 下载

2）安装

运行下载得到的 postgresql-13.3-2-windows-x64.exe，开始 PG 的安装：

（1）选择目标文件夹，如图 1-12 所示。

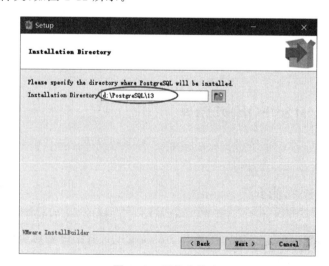

图 1-12 目标文件夹

(2)选择组件,如图1-13所示,这里全部选中。
(3)选择存放数据的文件夹,如图1-14所示。

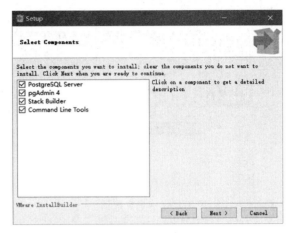

图1-13 选择组件

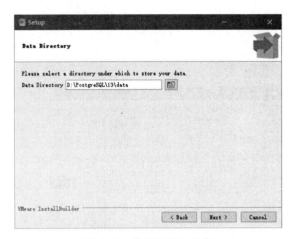

图1-14 选择数据文件夹

(4)设置数据库自带的默认管理员postgres的密码,如图1-15所示。
(5)设置数据库端口号,如图1-16所示。默认端口号是5432,建议使用默认值。

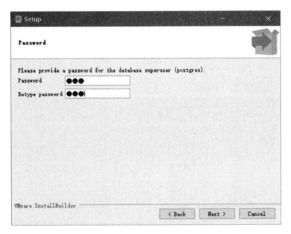

图1-15 设置密码

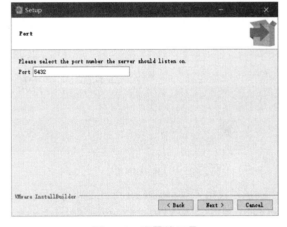

图1-16 设置端口号

后续步骤,都可以直接单击Next按钮,完成安装。最后一步弹出的Stack Builder向导窗口,用于下载并安装一些辅助工具,例如语言包、数据库驱动程序等。可以不勾选,直接单击Finish按钮,跳过这一步骤。

4. 创建数据库管理员admin

下面来创建一个数据库管理员admin。

(1)单击"开始"菜单PG中的pgAdmin4,这是PG自带的数据库管理工具。为方便起见,先将该工具的界面语言修改为中文。选择菜单File→Preferences→Miscellaneous→User language,然后在User下拉列表框中选择Chinese(Simplified),再单击Save按钮,如图1-17所示。

(2)重新打开pgAdmin4。在"登录/组角色"上右击,在弹出的快捷菜单中选择"创建"→"登录/

组角色",如图 1-18 所示。弹出创建窗口,在"常规"→"名称"中输入:admin,单击"定义",在密码框中输入密码,例如 007。再单击"权限",全选"是",如图 1-19 所示。最后,单击"保存"按钮即可。本书后续章节使用 admin 与数据库交互。

图 1-17　设置为中文

图 1-18　创建登录用户

图 1-19　设置权限

5. 创建数据库 tamsdb

在"数据库"上右击,在弹出的快捷菜单中选择"创建"→"数据库",如图 1-20 所示。在弹出对话框的"常规"→"数据库"后的文本框中输入数据库名:tamsdb,"所有者"选择 admin,如图 1-21 所示。单击"保存"按钮,数据库创建完毕。

图 1-20　创建数据库

图 1-21　定义数据库

提示：也可以使用MySQL或SQL Server等其他数据库产品，稍加修改，完全可以实现本书全部示例中的数据处理内容。感兴趣的读者，可以将此作为一种练习。

1.3.4 使用IntelliJ IDEA

1. IntelliJ IDEA 简介

IntelliJ IDEA是一款高效、友好的集成开发工具，堪称Java IDE（集成开发环境）的王者。效率化、智能化、多语言支持、智能代码补全等特性，使得IntelliJ IDEA在Java开发市场的占有率高达72%左右，使用者对IntelliJ IDEA的满意度高达98%左右。

除了受到个人开发者青睐外，IntelliJ IDEA也得到众多著名公司或机构的支持，例如谷歌、惠普、三星、大众汽车、美国航空航天局等。

2. 下载

可以到IntelliJ IDEA官网下载，官网提供只有30天试用期的旗舰版和免费的社区版。社区版在功能上有较多的限制。建议下载30天试用期的旗舰版，如图1-22所示。

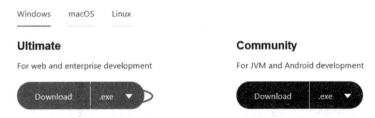

图1-22　IntelliJ IDEA下载

3. 安装

运行下载的IntelliJ IDEA安装程序，开始安装过程：
（1）更改目标文件夹为d:\JetBrains\IntelliJ IDEA，如图1-23所示。
（2）根据自己的需要，勾选若干自定义选项，如图1-24所示。

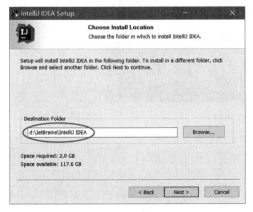

图1-23　选择目标文件夹

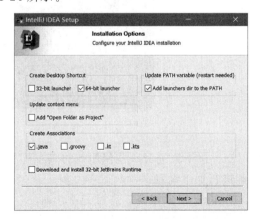

图1-24　定制选项

(3) 单击 Next 按钮,后续直接单击 Install 完成整个安装过程,再重启计算机。

4. 基本配置

(1) 启动 IntelliJ IDEA,勾选确认并接受用户协议,单击 Continue 按钮,选择不导入任何配置,在弹出的对话框中选择 Evaluate for free 进行 30 天试用,如图 1-25 所示,再单击 Evaluate→Continue 按钮,弹出欢迎主页,如图 1-26 所示。

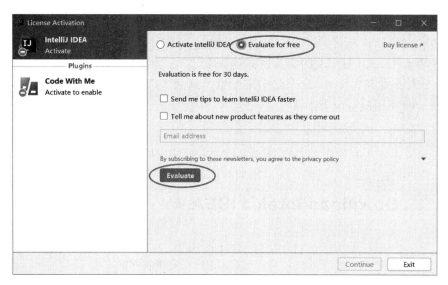

图 1-25 激活试用

图 1-26 欢迎主页

(2) 接下来进行一些使用前的基本配置。单击 Customize→All settings,弹出 Settings 对话框,在此进行如下设置:

◇ 设置界面外观。通过 Appearance & Behavior→Appearance→Theme,设置外观为 IntelliJ Light 风格。可根据个人喜好,设置为其他风格。

◇ 设置字符编码。通过 Editor→File Encodings，统一字符编码设置为 UTF-8，如图 1-27 所示，再单击 Apply 按钮。

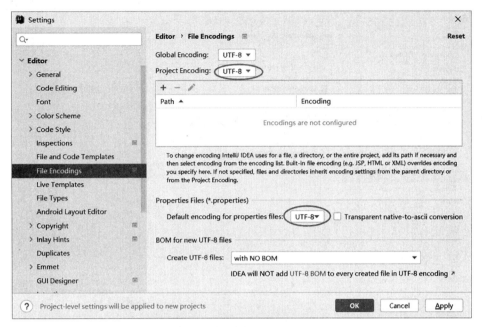

图 1-27　统一编码设为 UTF-8

◇ 自动编译项目。单击 Build，Execution，Deployment→Compiler 选项，勾选 Build project automatically，再单击 Apply 按钮，如图 1-28 所示。

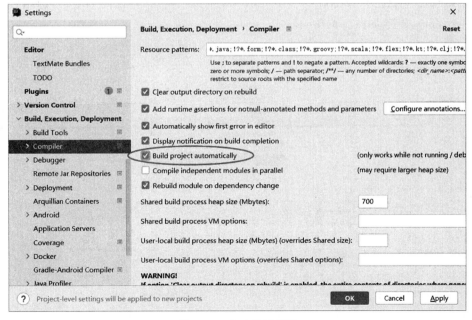

图 1-28　自动编译项目

◇ 配置应用服务器。单击 Build, Execution, Deployment→Application Servers 选项,再单击"+",选择 Tomcat Server,如图 1-29 所示。

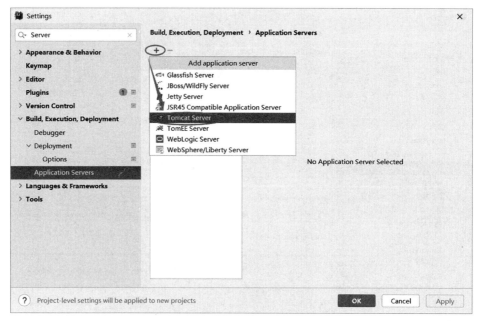

图 1-29　配置应用服务器

弹出 Tomcat Server 对话框,设置 Tomcat Home 为前面 1.3.2 节中安装好的 Tomcat 文件夹：D:\apache-tomcat-9.0.50,如图 1-30 所示。

◇ 设置默认浏览器。单击 Tools→Web Browsers 选项,根据自己的需要,勾选相应浏览器,再单击 Apply 按钮,如图 1-31 所示。

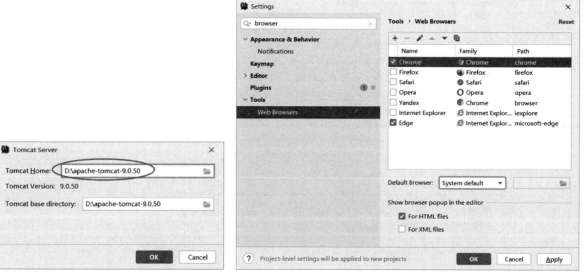

图 1-30　配置 Tomcat Home　　　　　　　图 1-31　设置浏览器

◇ 修改 HTML 页面模板。依次选择 Editor→File and Code Templates→HTML File 选项,如图 1-32 所示。将默认的英文 lang="en"修改为 lang="zh"简体中文,再单击 Apply 按钮。

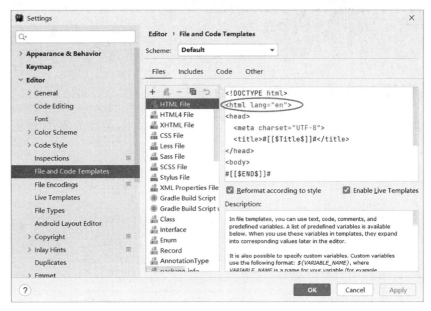

图 1-32　设置 HTML 页面语言

1.4　创建 Maven Web 站点

视频讲解

1.4.1　Maven 简介

Apache Maven 是 Java 项目构建、管理的工具,可以帮助开发者进行项目构建、依赖管理、项目描述信息管理、版本管理等,是构建工具领域的王者。

在开发 Java Web 项目的过程中,经常需要下载各种各样的资源。使用 Maven,就会方便很多。只需要告诉 Maven 需要什么,Maven 就会自动下载相应文件,非常方便。

1.4.2　修改资源仓库的镜像地址

默认情况下,Maven 会去我们所需资源的官方网站下载,例如国外的某个站点,很多时候下载速度可能非常缓慢甚至无法下载,非常不方便。好在国内一些厂商或大学提供了 Maven 资源仓库,例如阿里云、网易、清华大学、上海交通大学等。下面,我们设置 IntelliJ IDEA 的 Maven 资源仓库为阿里云仓库。用记事本打开 D:\JetBrains\IntelliJ IDEA\plugins\maven\lib\maven3\conf\settings.xml,在<mirrors></mirrors>中加入阿里云仓库地址:

```
<mirrors>
    <mirror>
        <id>alimaven</id>
```

```
                <mirrorOf> central </mirrorOf>
                <name> aliyun maven </name>
                <url> http://maven.aliyun.com/nexus/content/groups/public </url>
        </mirror>
</mirrors>
```

1.4.3 创建 Maven WebApp 站点项目

现在来创建一个 Maven Web 站点 chapter1。为了管理方便，可以先在 E 盘创建一个文件夹 myworks，用于存放本章及后续章节的 Web 站点项目。

(1) 单击欢迎主页(见图 1-26)的 New Project。选择 Maven→maven-archetype-webapp：RELEASE，再单击 Next 按钮，如图 1-33 所示。

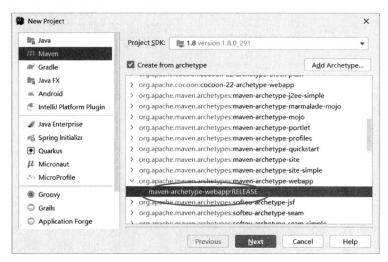

图 1-33　创建 Maven 项目

(2) 输入项目名称。还可以根据需要，设置项目辅助信息，例如 GroupId、ArtifactId、Version 等，如图 1-34 所示。

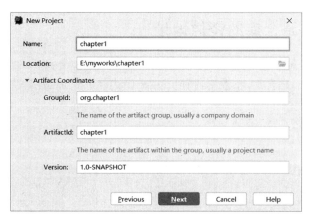

图 1-34　定义项目信息

（3）设置 Maven。这里直接采用 IntelliJ IDEA 提供的默认值，无须修改，如图 1-35 所示。单击 Finish 按钮，完成项目的创建。创建的 chapter1 项目结构，如图 1-36 所示。

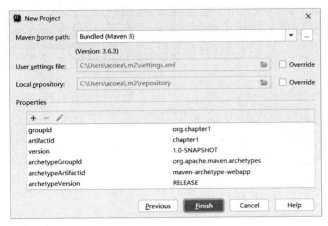

图 1-35　设置 Maven

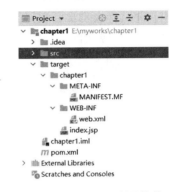

图 1-36　chapter1 项目结构

（4）为项目添加 Tomcat 服务器。单击 Add Configuration，弹出 Run/Debug Configurations 对话框。单击"＋"，找到并选择 Tomcat Server→Local，如图 1-37 所示。

（5）再单击 Deployment→Artifact。现在，需要定义 Java Web 项目的发布方式。此时有两种方式可以选择，如图 1-38 所示。

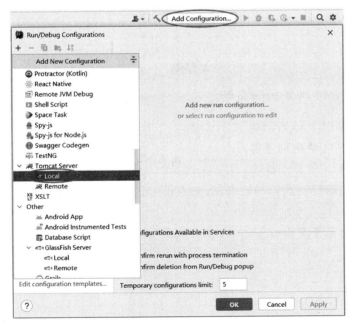

图 1-37　添加本地 Tomcat

图 1-38　选择发布方式

第一种方式是将 chapter1 项目打包压缩成 war 文件；第二种方式是 war 解压方式，也就是文件夹输出方式。两种方式都是可以的，只不过第一种方式正式发布到服务器更方便，因为只需要发布一个 war 压缩文件即可。这里选择第二种发布方式，单击 OK 按钮后，记得将 Application context 文本框中的输出路径修改为更简洁的"/chapter1"。

（6）单击 Server，修改 URL 以及其他一些选项，如图 1-39 所示。URL 就是 chapter1 站点的访问地址，默认的主页名是 index.jsp。On Update action 用于设置当项目内容发送变化时，自动更新编译 Java 类文件及资源文件。On frame deactivation 则用来设置在进行界面切换操作时的处理动作，这里仍然选择自动更新编译 Java 类文件及资源文件。这样选择的目的，主要是为了保证运行的是最新的项目资源文件。

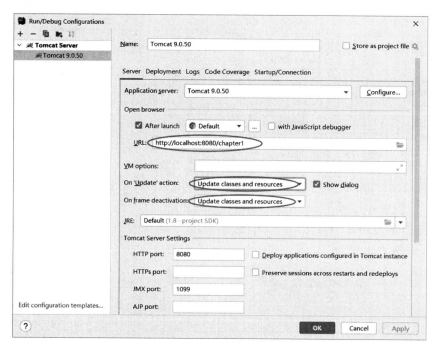

图 1-39　设置 Server 信息

（7）启动 chapter1 项目方法。单击如图 1-40 所示的三角形按钮，即可启动 Tomcat 服务器并自动在浏览器中打开 chapter1 站点，显示"Hello World!"字样。这个

图 1-40　项目运行按钮

三角形按钮右边的按钮，则采用调试方式运行站点。一般来说，站点没有最终编写完成前，建议用调试方式运行。

注意：启动项目前，请确保 Windows 服务里面的 Tomcat 不是处于"正在运行"状态。

1.4.4　修改项目编译版本

打开 pom.xml 文件，找到下面的代码，将 1.7 修改为 1.8。

```
<properties>
    <project.build.sourceEncoding>UTF-8</project.build.sourceEncoding>
    <maven.compiler.source>1.7</maven.compiler.source>
    <maven.compiler.target>1.7</maven.compiler.target>
</properties>
```

注意：后续章节创建的项目，都是基于1.8版本。

1.5 场景任务挑战——配置自己的站点

  1. 以自己姓名拼音为名，创建一个Maven Web站点，将主页内容修改为"欢迎访问某某的站点"字样。启动Tomcat服务器并打开该站点。

  2. 将Tomcat端口修改为80，修改站点配置。修改后，要保证第1步创建的站点，仍然能够正常打开。

HTML 基础

几乎所有的 Web 页面，都用到了 HTML。本章用不多的篇幅，主要介绍 HTML 基本结构、首部标签、正文标签，以及一些常用标签。熟练掌握 HTML，就掌握了一把通向 Web 页面世界的钥匙。

2.1 HTML 简介

HTML 是 Hyper Text Markup Language 超文本标记语言的简称。HTML 使用标记标签来描述网页，这些标记标签就是我们常说的 HTML 标签（HTML tag）。

HTML 标签是由尖括号包围的关键字，例如< html >。标签通常是成对出现的，比如 < html > 和 </html>，前者称为开始标签，后者称为结束标签。注意：有极少量 HTML 标签并非成对出现，而是单独出现的，例如换行标签< br/>。

我们常说的网页，实际上就是包含很多 HTML 标签的 HTML 文档。当我们在浏览器中打开 HTML 文档时，浏览器不会直接显示文档中的标签，而是解析这些标签，并显示相应的内容。例如，浏览器解析到< br/>标签，由于< br/>表示换行，就会在网页输出换行，呈现出换行效果。

HTML 非常容易学习！我们可以简单到用记事本编写 HTML 文档，或者用其他 IDE 工具，例如 Dreamweaver、IntelliJ IDEA 等。HTML 文档的扩展名是 html 或者 htm，建议使用 html 作为 HTML 文档的扩展名。

最新的 HTML 标准是 HTML5。HTML5 能够适应计算机、手机等不同类别设备，提供了很多新元素，极大地丰富了 Web 内容的展现。本章所介绍的内容里面已经包含了 HTML5 的部分内容。

2.2 HTML 文档基本结构

HTML 文档的基本结构如下：

```html
<!DOCTYPE html>                    ①
<html lang="zh">                   ②
<head>                             ③
    <meta charset="UTF-8">
    <title>小萌的小窝</title>
</head>
<body>                             ④
小萌的站点，欢迎访问！
</body>
</html>
```

说明：HTML 文档由以下四部分组成。

① 类型声明。这句必须放在 HTML 文档的第一行，说明该文档为 HTML 文档。

② 语言类型。声明页面中的主要语言是简体中文(zh)，此外还可以用 en(英文)、de(德文)、ja(日文)等。声明语言类型，对搜索引擎和浏览器能够提供搜索或解释方面的帮助。

③ 头部。这里包含的内容为浏览器提供一些指示信息。

④ 主体。这里就是网页需要显示的内容。

2.3 头部和主体标签

2.3.1 头部标签<head>

<head>标签是用来定义头部元素的容器。<head>内的元素可以是 JavaScript 脚本、CSS 样式表、Meta 元数据信息、页面标题等。例如：

```html
<head>
    <meta charset="UTF-8">
    <title>教务辅助管理系统--主页</title>
    <link href="css/home.css" rel="stylesheet" type="text/css">
    <script src="module/login.js"></script>
</head>
```

这个<head>标签内，定义了网页使用的字符集为 UTF-8，该网页在浏览器中显示的标题是"教务辅助管理系统--主页"，并告诉浏览器该页面使用的样式表是 css 文件夹下的 home.css，同时包含了 module 文件夹下的 JavaScript 脚本文件 login.js 以供程序调用。关于 CSS 和 JavaScript，我们将在第 3 章、第 4 章分别讨论。

2.3.2 主体标签<body>

主体标签<body>定义了文档的主体内容，可以将文字、图片、超链接、表格、视频等需要的内容放置其中。例如：

```
< body >
    小萌的站点,欢迎访问< img src = "image/at.png">
    < a href = "https://www.hust.edu.cn">华中科技大学</a>
</body >
```

上面页面中,将会显示欢迎文字、一张图片 at.png、导航到华中科技大学的超链接。

2.4 其他常用标签

HTML 标签比较多,这里仅介绍实际中用得比较多的一些标签。

2.4.1 链接分段与换行

1. 链接

链接标签< a >在网页中用得相当频繁。基本格式如下:

```
< a href = "地址或页面名" target = "目标窗口" title = "提示文本">链接名</a>
```

◇ href 属性定义了链接目的地。href 的值,可以是类似于 https://www.hust.edu.cn/这样的绝对地址,也可以是相对地址,例如当前网站的某个页面:login.html。
◇ target 属性用来设置打开链接页面的方式。用得比较多的是:
target = "_self",在当前页面打开;
target = "_blank",在新窗口打开;
target = "_parent",在上一级页面打开。
◇ title 属性用于给链接设置提示文本,例如 title = "招聘专业:计算机软件 "。

示例:三个不同的链接。

```
< a href = "https://www.hust.edu.cn" target = "_blank">华中科技大学</a>
< a href = "https://www.whu.edu.cn" target = "_blank">武汉大学</a>
< a href = "save.html" target = "_self" title = "保存数据到数据库">保存</a>
```

2. 分段与换行

分段标签< p >……</p>。
换行标签< br/>。
示例:

```
<! DOCTYPE html >
< html lang = "zh">
< head >
    < meta charset = "UTF - 8">
    < title>示例</title>
</head >
```

```
<body>
<p>实现中华民族伟大复兴是中华民族近代以来最伟大的梦想</p><p>自力更生、发愤图强、解放思想、锐意进取</p><p>千秋伟业、征程如虹,<br/>百年奋斗、气壮山河。</p>
</body>
</html>
```

上面代码用<p>定义了三段文字,而最后一段文字,则在"征程如虹,"后换行显示。在浏览器中的显示效果如图 2-1 所示。

图 2-1 分段换行效果

提示:为简便起见,后续示例,将只提供主要部分的代码,省略粗体部分的代码。

2.4.2 表格和列表

1. 表格< table >

标签< table >用来定义 HTML 表格元素,非常适合需要用规整形式展示数据的场景。表格常用表示法如下。

- ◇ thead、tbody、tfoot:分别表示表头、主体、表尾。
- ◇ caption、th:分别表示表格标题、表头中的数据。
- ◇ tr、td:分别表示行、单元格。
- ◇ colspan、rowspan:前者定义单元格可横跨的列数,后者定义单元格可横跨的行数。
- ◇ width、height、border:分别定义表格宽度、高度、边框。

来看下面的表格:

表格示例

编号	汇总	
01	管理学院	
0101	信息管理	185
0102	电子商务	
		总计:185

这个表格 5 行 3 列,表格标题为"表格示例",表头为"编号""汇总"。单元格"汇总""管理学院"横跨 2 列,单元格"185"横跨 2 行。要在网页中显示这张表格,用 table 标签,从左到右、从上到下,依序写完每个单元格即可实现,完整代码如下:

```
<table border="1" width="25%">
    <caption>表格示例</caption>
    <thead>
        <tr>
            <th>编号</th><th colspan="2">汇总</th>
        </tr>
```

```html
        </thead>
        <tbody>
            <tr>
                <td>01</td><td colspan="2">管理学院</td>
            </tr>
            <tr>
                <td>0101</td><td>信息管理</td>
                <td rowspan="2">185</td>
            </tr>
            <tr>
                <td>0102</td><td>电子商务</td>
            </tr>
        </tbody>
        <tfoot>
            <tr>
                <td colspan="3" align="right">总计：185</td><!-- 右对齐 -->
            </tr>
        </tfoot>
</table>
```

上面表格的宽度 width 用百分比形式来定义，也可以用具体的数值，例如 width="350" 设置表格宽度。像表格宽度、边框、对齐等内容，以后都采用 CSS 修饰更好！

2. 列表

有两种形式的列表：有序列表 `<ol>`、无序列表 `<ul>`，列表中的列表项用 `<li>` 表示。来看下面的代码：

```html
<ol start="10">
    <li>管理类</li>
    <ul>
        <li>工商管理</li>
        <li>运营管理</li>
    </ul>
    <li>计算机类</li>
    <ul>
        <li>计算机网络</li>
        <li>Java Web</li>
        <li>Java 程序设计</li>
    </ul>
    <li>经济类</li>
    <ul>
        <li>微观经济学</li>
        <li>统计学原理</li>
        <li>计量经济学</li>
    </ul>
</ol>
```

上面的代码用有序列表定义了三类书籍：管理类、计算机类、经济类，并用 start 属性定义了列表的起始值为 10，而每类下面的具体书名，则用无序列表展示。在浏览器中的运行效果，如图 2-2 所示。

2.4.3 层标签和组合标签

1. 层标签< div >

层< div >很适合用来进行网页的布局，例如将网页划分成几个区，每个区显示不同的内容。层< div >也充当着容器的作用，可以在 div 中放置其他 HTML 元素，例如文字、图片、表格等。

图 2-3 叠加的层

层< div >有一个重要属性 style，可用来对 div 进行外观修饰。style 属性是一个全局属性。所谓全局属性，就是绝大部分 HTML 元素都支持这个属性，例如前面学过的 body、p、table、ol、ul 等都有 style 属性。

来看一个示例。图 2-3 用两个层上下叠加在一起，并做了样式的修饰处理：底部的层为圆角矩形，靠上的层则是椭圆形。圆角矩形层是浅红色背景，并有"你好，Java Web!"字样，椭圆形层是白色背景，并放置了一张 png 图片。

下面是这两个叠加层的代码实现：

```
< div style = "position: absolute;
              left: 226px;
              top: 125px;
              width: 118px;
              height: 116px;
              font-size:12px;
              border-radius: 6%;
              z-index: 1;
              background-color: #ffcccc;">
     你好，Java Web!
</div>
< div style = "position: absolute;
              left: 188px;
              top: 156px;
              width: 186px;
              height: 62px;
              border-radius: 48%;
              z-index: 2;
              text-align:center;
              background-color: #fff;">
     < img src = "m0.png" width = 48 height = 52/>
</div>
```

上面的代码，印证了层作为容器确实可以非常方便地实现网页内容的布局或划分。代码的关键其实是两个层 style 属性的内容。这些内容属于第 3 章 CSS 的范畴，这里无须完全理解，不过我们还是可以从单词本身管中窥豹，例如 left、top 定义了层的位置，width、height 定义了层的大小等。

看起来,style 的写法比较烦琐。确实如此!这并不是推荐的写法,这里只是为了展示层的应用才这样做。学习完第 3 章以后,我们就会采用更好的写法。

2. 组合标签<span>

组合标签<span>用于组合 HTML 文档中的各种元素,常常作为文本或其他标签元素的容器。下面的示例,用 span 组合显示蓝色诗句和一张图片:

```
<span style = "color:#00F">
    问渠那得清如许,为有源头活水来
    <img src = "image/m9.gif" width = 48 height = 52/>
</span>
```

2.4.4 图像和媒体

1. 图像

1) img

<img>标签定义了图像元素的显示。<img>标签的使用格式示例如下:

```
<img src = "image/query.png" alt = "查询" width = "32" height = "32" title = "开始查询">
```

- ◇ src 属性指定图像的来源。这里使用了 image 文件夹下的 query.png 图片。
- ◇ alt 属性为图像定义了预备的可替换文本。一旦浏览器无法载入图像时,就会在原本显示图像的位置显示替换文本,例如"查询"二字。
- ◇ width、height 属性分别定义了图像的宽度、高度,可以使用具体数值或者百分数形式。尽管这两个属性都可以忽略不写,此时浏览器会视情况缩放显示图像,但显式地指明 width、height 是一个好习惯,有利于浏览器更快地加载图像。
- ◇ title 属性用来指定图像的提示文本。当鼠标指向图像时,会显示"开始查询"字样。

2) picture

与<img>不同,<picture>能够基于定义的图像集,根据浏览器窗口大小,自动匹配最适合显示的图片,因而更具有灵活性。其使用格式示例如下:

```
<picture>
    <source media = "(min - width:116px)" srcset = "image/m0.gif">
    <source media = "(min - width:48px)" srcset = "image/m1.gif">
    <img src = "image/m2.gif" style = "width:auto;">
</picture>
```

根据浏览器窗口大小,自动匹配与<source>标签中所定义宽度最适合的图像。<img>标签通常作为<picture>标签的最后一个元素,如果上面<source>定义的图像都不匹配,则作为备选项显示。

2. 媒体

标签<video>用来定义视频播放,<audio>则用来定义音频播放。两个标签的使用形式相差不大。

1) 视频播放示例

```
<video width="640" height="480" controls="controls" src="video/a.mp4" type="video/mp4" autoplay="autoplay">
您的浏览器不支持视频播放标记!
</video>
```

◇ controls 属性指定是否显示播放控制的控件,这里设置为显示播放控件。
◇ src 属性指定了视频源为 video 文件夹下的 a.mp4 文件,type 属性则指定了视频类型为 mp4。
◇ autoplay 属性用来指示视频在就绪后立即播放。
◇ 一旦用户浏览器不支持<video>标签,则在网页上显示"您的浏览器不支持视频播放标记!"的提示文字。

2) 音频播放示例

```
<audio src="audio/myf.mp3" controls="controls" loop="loop">
    您的浏览器不支持音频播放标记!
</audio>
```

这里用 loop 属性指定该音频反复播放。

2.4.5 对话框

标签<dialog>用于显示一个对话框。这个标签的使用还是比较简单的:

```
<dialog open>数据保存成功!</dialog>
```

标签的 open 属性用来控制对话框的显示或不可见。不过,可以更进一步对<dialog>做些额外处理,使对话框更为美观、实用,如图 2-4 所示:

相应 HTML 代码如下:

图 2-4 对话框

```
<dialog style="text-align:center;border:1px solid #ccc;" open>
    <span>
        <img src="image/note.png" width=32 height=32/>数据保存成功!
    </span><br/>
    <button id="confirmBtn">确  定</button>
</dialog>
```

代码中添加了一个<span>标签来显示灯泡图标和提示文字,并使用<button>显示一个"确定"按钮。<button>标签下一节就会学习到。" "是一个 HTML 编码,代表全角空格,类似的还有" "表示半角空格。当我们学习了第 4 章 JavaScript 的内容后,就可以更灵活地控制这个对话框,例如某个操作完成后,弹出对话框;单击"确定"按钮,关闭对话框。

2.4.6 表单及表单元素

1. 表单

表单<form>是用来收集站点访问者信息的集合,例如用户登录信息、注册信息以及其他各种提交信息。<form>基本格式如下:

```
<form name = "form1" method = "get | post" action = "处理页面" onSubmit = "处理程序">
    ...
</form>
```

- ◇ name 属性设置表单名称。
- ◇ method 属性定义提交表单数据时使用的 HTTP 方式:get 或 post。①get 方式,是默认的数据提交方式,被提交的数据附加在负责处理页面(例如 tj.html)的地址后面。假设提交的数据只有两个,username 值为 aaa,password 值为 007,则 get 提交数据地址栏 URL 形式为:tj.html?username=aaa&password=007。一旦提交的数据很多,地址栏显示的 URL 就会很长且用户可以看到被提交的数据,安全性较弱,所以 get 方式常用于提交数据较少或者不包含敏感数据(例如密码)的场合。②post 方式提交的数据,数据被封装在请求体里面,在页面地址栏是不会显示的,安全性更好。get 请求浏览器将 HTTP 请求头和数据一并发送,而 post 请求多数浏览器会先发送 HTTP 请求头信息,服务器响应后再视情况发送数据,更严谨些。实际上,无论少量或较多数据,post 都能胜任,所以采用 post 方式提交数据的居多。
- ◇ action 属性设置表单提交后负责处理的页面,例如 action = "tj.html"。如果没有指定,则使用当前页面。
- ◇ onSubmit 是事件处理程序,用来定义单击表单的提交按钮时的处理动作,例如 onSubmit = "check()"。利用 onSubmit,就可以在表单提交之前通过 check()进行一些必要的检查工作,例如用户注册时检查用户名、密码是否按照要求填写了。

表单通常需要配合表单元素来进行各种处理。

2. 表单元素

1) input

常用于各种输入场合,是一种输入型表单控件,包括输入具体的数据,也包括鼠标单击这种特殊"输入"。基本格式:

```
<input type = "类型" name = "名称" value = "值">
```

- ◇ type 决定了 input 的类型。类型值比较多,比较常见的有:text(文本框)、password(密码框)、checkbox(复选框)、radio(单选按钮)、hidden(隐藏域)、submit(提交按钮)、reset(重置按钮)、button(常规按钮)、email(电子邮件)、url(网址)、number(数字)、range(范围)、date(日期),等等。
- ◇ name 定义 input 的名称。
- ◇ value 代表 input 的值,这个"值"可能是你输入的密码,也可能是按钮上面的文字。

示例1：登录表单。

```
< form name = "form1" method = "post" action = "tj.jsp">
    用户名< input type = "text" name = "usr"/>
    密 码< input type = "password" name = "pwd"/>
    < input type = "submit" value = "登录"/>< input type = "reset" value = "重填"/>
</form >
```

这里需要注意的是：登录按钮的类型为submit，这种类型专门用于提交表单，而reset按钮则会将用户输入的内容清空。页面在浏览器中的运行效果如图2-5所示。

示例2：限制数字输入。

```
< form name = "form2" method = "post" action = "num.html">
    请输入分数:< input type = "number" name = "score" min = "1" max = "100"/>
    < input type = "submit" value = "提交"/>< input type = "reset" value = "重填"/>
</form >
```

运行效果如图2-6所示。

图 2-5　登录表单　　　　　　　图 2-6　录入分数

示例3：日期选择器。

```
< form name = "form3" method = "post" action = "reg.html">
    注册日期: < input type = "date" name = "registDate"/>
    < input type = "submit" value = "提交"/>< input type = "reset" value = "重置"/>
</form >
```

运行效果如图2-7所示。

2) datalist

数据选项列表标签。为了方便数据输入，可以在用户输入数据时弹出预先定义好的下拉选项列表，这样就能够实现快速选择到需要的数据。数据选项列表datalist基本格式如下：

```
< datalist id = "识别号">
    < option value = "值1">
    < option value = "值2">
    ...
</datalist >
```

图 2-7　日期选择器

图 2-8　下拉选项列表

下面的示例，提供了五个可供查询的专业名称。当用户在输入框输入时，会自动弹出下拉选项列表，并能够根据用户输入的文字，自动过滤，缩小选择范围，非常方便快捷。运行效果如图2-8所示。

```html
<form name="form4" method="post" action="query.jsp">
    <input list="majors" name="major">
    <datalist id="majors">
        <option value="信息管理与信息系统">
        <option value="电子商务">
        <option value="集成电路">
        <option value="软件工程">
        <option value="人工智能">
    </datalist>
    <input type="submit" value="开始查询"/><input type="reset" value="重置数据"/>
</form>
```

注意：<input>标签的 list 属性值必须与<datalist>标签的 id 属性值相同。

3) select

<select>标签定义下拉选择框。来看下面的示例：

```html
<select name="major">
    <option value="0701">信息管理与信息系统</option>
    <option value="0702">电子商务</option>
    <option value="0703" selected>集成电路</option>
    <option value="0704">软件工程</option>
    <option value="0705">人工智能</option>
</select>
```

<option>用于定义选择项，value 属性代表选择项被选中时的值，例如如果选中"电子商务"，则当前<select>选择项的值是"0702"。另外，selected 设置该选择项为页面打开时的默认选中项。

4) button

该标签用于定义按钮，形式比较简单，例如：

```html
<button type="button" onclick="alert('你好,Java Web!')">测试按钮</button>
```

单击该按钮时，会弹出显示"你好，Java Web!"字样的对话框。<button>里面也可以添加其他内容，例如图片，这样的话就会显示一个图片＋文字的按钮，更为美观。

5) textarea

<textarea>标签是一种用于输入多行文本的控件，非常适合用于个人简介、工作经历、公司介绍、奖惩情况、备注等较多内容场景的输入。示例应用：

```html
<textarea name="intro" rows="6" cols="40" placeholder="输入个人简介……"></textarea>
```

placeholder 属性定义了文本区域的简短提示。

2.4.7 模板

<template>标签用于设置需要对用户隐藏的内容。来看示例:

```
<template>
    <img src = "image/m0.gif" width = "32" height = "32">
    <button type = "button" id = "resetBtn">数据重置</button>
</template>
```

<template>标签里面的图片和按钮,在网页打开时是不会显示的。不过,无须担心,我们学习了 JavaScript 脚本后,就可以根据页面处理的需要来呈现<template>中隐藏的内容。

2.4.8 内联框架

内联框架<iframe>标签有助于实现在一个网页里面呈现另外一个网页的内容。下面通过例子来学习这个标签的使用,假如有两个页面 a.html、b.html:

1) a.html 的主要内容

```
<body>
你好,Java Web!<br/>
<iframe src = "b.html" frameborder = "0" width = "116" height = "145" scrolling = "no">
    您的浏览器不支持内联框架
</iframe><br/>
一直在路上……
</body>
```

在上面的代码中,frameborder 设置内联框架为无边框,scrolling = "no"设置不显示滚动条。<iframe>内联框架的内容来自 b.html。

2) b.html 主要内容

```
<body>
<img src = "image/gril.png" width = "116" height = "106">
努力前行
</body>
```

在浏览器中打开页面 a.html,显示效果如图 2-9 所示。显然,a.html 中包含了 b.html 页面的内容。

图 2-9 运行效果

2.5 事件

事件是一种触发动作,例如当用户在按钮上单击鼠标时,触发 onclick 事件。HTML 事件划分成五大类:Window(窗口)、Form(表单)、Keyboard(键盘)、Mouse(鼠标)和 Media(媒体)。表 2-1 只是列出了一些常用事件。

表 2-1　HTML 常用事件

事件名	类别	说　　明
onload	Window	页面结束加载之后触发
onresize	Window	浏览器窗口被调整大小时触发
onblur	Form	失去焦点时触发
onchange	Form	值被改变时触发
onfocus	Form	获得焦点时触发
onselect	Form	文本被选中后触发
onsubmit	Form	提交表单时触发
onkeydown	Keyboard	按下按键时触发
onkeyup	Keyboard	释放按键时触发
onclick	Mouse	单击鼠标时触发
onmousemove	Mouse	鼠标移动到元素上时触发
onmouseout	Mouse	鼠标移出元素时触发
onplaying	Media	播放时触发
onpause	Media	暂停时触发

来看下面的示例：

```
< body onload = "alert('欢迎访问本站点！')">
    < button type = "button" onclick = "alert('查询学生成绩…')">
        < img src = "image/search.png" width = "28" height = "28"><br/>成绩查询
    </button>
</body>
```

打开该页面，首先自动弹出"欢迎访问本站点！"对话框。当单击"成绩查询"按钮时，会弹出"查询学生成绩…"对话框，如图 2-10 所示。这里的"成绩查询"按钮，是一个图片与文字组合的按钮。

图 2-10　事件对话框

2.6　场景应用示例——显示字符映射表图标

视频讲解

2.6.1　应用需求

我们来做一个稍有趣味的示例：显示 Windows 系统自带的字符映射表图标。字符映射表是

Windows 自带的系统工具,提供了很多无法通过键盘输入的特殊字符,例如图 2-11 中的小喇叭。

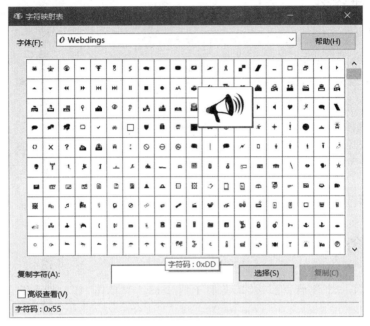

图 2-11 字符映射表

从字符映射表中挑选四个图标:喇叭、扩音器、蜘蛛、聊天,并在网页中显示出来,实际效果如图 2-12 所示。

图 2-12 字符映射图标

2.6.2 实现思路

那么如何实现呢?具体步骤如下:

(1) 确认图标所在的字体名称,例如由图 2-11 可知:小喇叭图标的字体是 Webdings。实际上,我们这个例子选取的图标都来自 Webdings 字体下的图标。大家也可以看看其他字体下面都有哪些图标,例如 Wingdings 字体。

(2) 获得当前图标的十六进制字符码。由图 2-11 左下角数字可知:小喇叭图标的字符码是 0x55。

(3) 计算出对应的十进制码。可利用 Windows 自带的计算器,计算后可得:小喇叭图标字符码的十进制数字是 85,然后用"U"表示小喇叭图标。

(4) 依次处理其他三个字符映射图标。

2.6.3 具体实现

下面是具体代码：

```
<body>
    <p style = "font-family:Webdings;
                font-size:14px;
                color:#00F;">
        <span>&#85;&#177;&#33;&#94;</span>
    </p>
</body>
```

提示：不是所有字符映射图标都能在网页中显示！

2.7 场景任务挑战——注册与叠加的层

1. 利用表单制作一个注册页面，效果类似于图2-13。当单击"提交注册"按钮时，弹出"开始注册……"对话框。
2. 试着将2.4.3节中层标签的示例（见图2-3）修改成如图2-14所示的效果。

图2-13 注册页面

图2-14 叠加的层

第3章 CSS(层叠样式表)

美轮美奂的网页,往往容易给人留下深刻印象,CSS 就是网页的装饰师。本章主要介绍 CSS 基本格式、应用方式、选择器、常见样式设置方法等内容。熟练掌握 CSS,是 Web 前端开发必备的技能。

3.1 CSS 简介

CSS 全称为 Cascading Style Sheets,层叠样式表,是万维网联盟 W3C 制定的一种标准。CSS 描述了如何在页面上显示 HTML 的各种元素,可以方便地控制网页的样式和布局。

CSS3 是 CSS 的升级版本,从多个方面进行了改进和扩充。本章所介绍的内容里面,已经包含了 CSS3 的部分内容。

3.2 CSS 基础

3.2.1 CSS 基本格式

CSS 由选择器和声明两大部分组成,以下面的 CSS 为例:

```
<style type="text/css">
    div {
        color:#00f;
        font-size:12px;
    }
</style>
```

◇ <style>:为页面定义 CSS 样式信息。
◇ div:选择器。选择器的意思就是选取需要设置样式的 HTML 元素,这里表示该 CSS 是对 div 层进行修饰。

◇ {}：大括号。定义了CSS声明块的内容。声明块包含了一条或多条用分号分隔的声明。这里包含了关于颜色、字体大小的两条声明。

◇ 声明。声明是具体用来对网页的各种元素进行修饰的部分。每条声明由"属性名：属性值"数据对组成，多个声明之间用冒号分隔，例如"color:♯00f;"中color为属性名，♯00f为属性值。

上面示例的CSS，控制页面中所有div层的文本颜色为蓝色，字体大小为12px。

CSS的颜色有下面三种表示方法。

(1) 用单词表示，例如blue表示蓝色，但单词只能表示有限数目的颜色。

(2) 采用十六进制表示，格式：♯RGB。RGB分别表示红、绿、蓝三原色，取值范围都是00～FF，大小写均可，例如♯0000FF(蓝色，可简写为♯00F)、♯F00(红色)、♯7D7DFF(浅蓝色)。

(3) 用RGB函数表示，例如红色rgb(255,0,0)。

3.2.2 应用方式

如果要用CSS进行修饰，一般有三种方式。

1. "标签名"作为选择器

```css
<style type="text/css">
    div {
        width:125px;
        border: 1px dotted ♯00f;
    }
</style>
```

页面代码：

```html
<span>观书有感</span>
<div>半亩方塘一鉴开</div>
<div>天光云影共徘徊</div>
```

这种方式下，样式会影响页面上所有<div>标签，即两句诗所在层都具有同样的样式：长度125px，边框为蓝色点线，线宽1px，但<span>不具有这种样式。

2. "♯自定义名"作为选择器

```css
<style type="text/css">
    ♯poem1 {
        width:125px;
        border: 1px solid ♯f00;
    }
    ♯poem2 {
        width:125px;
        border: 1px dotted ♯00f;
    }
</style>
```

页面代码：

```
<span>观书有感</span>
<div id="poem1">半亩方塘一鉴开</div>
<div id="poem2">天光云影共徘徊</div>
```

这种方式下，样式会影响页面上所有id值与选择器名称相同的页面元素。所以，"半亩方塘一鉴开"的边框为红色实线、线宽1px，而"天光云影共徘徊"的边框为蓝色点线、线宽1px。

3. ".自定义名"作为选择器

```
<style type="text/css">
    div {
        width: 130px;
        text-align: center;
        padding: 1px;
        background-color: #fff;
        display: table-cell;
    }
    img{
        vertical-align:middle;              ①
    }
    #title {
        color: #f00;
    }
    .poem {                                 ②
        border: 1px dotted #00f;
    }
    div.poem:nth-child(odd) {               ③
        background-color: #e0e0f3;
    }
    div.poem:nth-child(even) {              ④
        background-color: #e8f1fb;
    }
</style>
```

页面代码：

```
<div>
    <img src="image/note.png" width="16" height="16">
    <span id="title">观书有感</span>
    <img src="image/note.png" width="16" height="16">
    <div class="poem">半亩方塘一鉴开</div>
    <div class="poem">
        <span>天光云影共徘徊</span>
    </div>
    <div class="poem">问渠那得清如许</div>
    <div class="poem">为有源头活水来</div>
</div>
```

说明：

① 图片与文字在垂直方向居中对齐。

② 这种方式下，样式会影响页面上所有 class 值与选择器名称相同的页面元素。所以，四句诗的边框均为蓝色点线、线宽 1px，而诗名"观书有感"为红色字、无边框。

③ class="poem"的所有<div>中，奇数位置的<div>背景色为♯e0e0f3。

④ class="poem"的所有<div>中，偶数位置的<div>背景色为♯e8f1fb。

即③④样式设置了四句诗的奇偶行背景颜色的交替变化，如图 3-1 所示。

图 3-1 奇偶行变色

结合上面三种应用方式，再拓展介绍几种选择器用法：

◇ div.poem：选择 class="poem"的所有<div>元素，例如 div.poem{font-size:12px;}，下面几种与此类似。

◇ div span：选择<div>内部的<span>元素。

◇ img,span：选择所有<img>、<span>元素。

◇ div[class^="poem"]：选择 class 属性值以"poem"开头的<div>。

◇ img[src$=".png"]：选择 src 属性值以".png"结尾的<img>。

◇ img[src~="on"]：选择 src 属性值中包含"on"的<img>。

◇ img+span：选择紧跟<img>的首个<span>。

3.3 CSS 样式设置

熟悉了 CSS 基本语法和应用方式后，就可以进一步学习 CSS 的各种样式设置。CSS 样式设置非常丰富，下面选择一些常用样式设置进行介绍。

3.3.1 文本

通过表 3-1 来学习一些常见的文本设置。

表 3-1 文本设置

名称	说明	举例（作用）
color	颜色	color:♯00f;（蓝色文字）
font-family	字体	font-family:Wingdings;（文鼎字体）
font-style	字体样式	font-style:italic;（斜体字）
font-weight	字体粗细	font-weight:bold;（文字加粗）
font-size	字体大小	font-size:1em;（字体大小为 16 像素）
text-align	水平对齐方式	text-align:center;（水平方向居中）
vertical-align	垂直对齐方式	vertical-align:middle;（垂直方向居中）
text-decoration	装饰线	text-decoration:underline;（文字有下画线）
text-transform	文本转换	text-transform: uppercase;（转换为大写）
letter-spacing	字符间距	letter-spacing:−5px;（字符间距 5 像素）
word-spacing	单词间距	word-spacing:5px;（单词间距 5 像素）
line-height	行间距	line-height:1.5;（增大行间距）
text-shadow	文字阴影	text-shadow:1px 1px 1px ♯ccc;（灰色阴影）

示例：用红色加粗字体突出显示部分文字。

```
<style>
    p.msg span {
        font-weight: bold;
        color: #f00;
        font-size: 1.2em;
        text-shadow: 2px 2px 3px #ccc;
    }
</style>
<div>
    <p class = "msg">
        <span>友情提示：</span>
        优惠活动截止到今天！
    </p>
</div>
```

这里仅设置"友情提示："为红色、粗体、阴影效果，如图 3-2 所示。

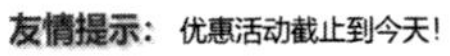

图 3-2　文本样式

3.3.2　背景

背景主要包括背景颜色、透明度、背景图片等。表 3-2 列出一些常见设置。

表 3-2　背景设置

名　称	说　明	举例（作用）
background-color	背景色	background-color:#0f0;（绿色背景）
opacity	透明度	opacity:0.5;（半透明）
background-image	背景图像	background-image:url("image/bg.png");（背景图像设置为 image 文件夹下的 bg.png）
background-repeat	重复图像	background-repeat:repeat-x;（水平方向拉伸图像） background-repeat:repeat-y;（垂直方向拉伸图像） background-repeat:no-repeat;（不拉伸图像）
background-position	图像位置	background-position:right top;（右上角）
background-attachment	背景图像固定方式	background-attachment:fixed;（固定图像） background-attachment:scroll;（可滚动图像）
background-size	图像大小	background-size:60px 35px;（图像大小 60×35 像素）

示例：将上面的例子稍微修改一下，加入背景图片。
其他部分不变，只需要在<style>中加入：

```
div{
    width:272px;
    height:75px;
    background-image:url("image/bg.png");
    background-size:272px 75px;
}
```

运行效果如图 3-3 所示。

3.3.3 边框和边距

图 3-3 背景图片

边框属性定义了边框的颜色、宽度、形状等，边距则定义了元素与边框的距离。表 3-3 列出了常见边框和边距属性。

表 3-3 边框和边距

名 称	说 明	举例(作用)
border-style	边框类型	border-style:1px dotted #00f;(蓝色点线边框,线宽 1px。边框类型还可以是 solid 实线、dashed 虚线、double 双线、none 无边框等)
border-width	边框宽度	border-width:1em 0.5em 1em 0.5em;(上右下左边框的宽度)
border-color	边框颜色	border-color:#f00 #00f #f00 #00f;(上右下左边框的颜色)
border-radius	圆角边框	border-radius:4px;(也可用百分数表示)
margin	边距	margin:5px 3px 5px 3px;(上右下左边距,可以分别用 margin-top、margin-right、margin-bottom、margin-left 代替)

示例：将上例中的<div>加上边框"上、下边框为红色点线，左右无边框"。
只需要在<style>的 div 选择器中加入以下代码：

```
border:1px dotted;
border-width: 1px 0px;
border-color: #f00;
```

3.3.4 定位溢出和浮动

position 定位设置元素的位置或前后顺序，overflow 用于设置在内容较多无法容纳时的处理方式，float 则指明了页面元素如何浮动。表 3-4 列出了定位、溢出和浮动的常见用法。

表 3-4 定位、溢出和浮动

名 称	说 明	举例(作用)
position	定位	position:relative;(相对定位,常见的还有 fixed、absolute 等)
z-index	前后顺序	z-index:−1;(数值越大越靠前)
overflow	溢出	overflow:hidden;(隐藏超出内容,还可以设置为 scroll、auto 等)
float	浮动	float:left(处于左侧)

示例：将各自包含一张图片的两个不同的层<div>，分别放置在背景层的左上角、右上角，如图 3-4 所示。

图 3-4 定位

代码如下:

```
<style>
    div {
        margin: auto;                                ①
        width: 260px;
        height: 75px;
        background-image: url("image/bg.png");       ②
        background-size: 260px 75px;
    }
    .left, .right {
        position: relative;                          ③
        width: 28px;
        height: 28px;
    }
    .left {
        float: left;                                 ④
    }
    .right {
        float: right;                                ⑤
    }
</style>
<div>
    <div class="left">
        <img src="image/db.png" width="28" height="28">
    </div>
    <div class="right">
        <img src="image/statistics.png" width="28" height="28">
    </div>
</div>
```

说明:

① 最外层<div>自动处于屏幕中央。
② 设置背景图片为 image/bg.png,请使用 url()函数指明图片位置。
③ 相同的样式放在一起定义,位置定义为相对于最外层<div>。
④ 浮动在最外层<div>的左边。
⑤ 浮动在最外层<div>的右边。

3.3.5 伪类和伪元素

1. 伪类

伪类常用于定义页面元素的特殊状态,例如鼠标悬停时改变颜色。表 3-5 列出一些常用的伪类。

表 3-5 伪类

名称	说明	举例(作用)
:active	活动的	a:active(活动链接)
:checked	被选中的	input:checked(被选中的复选框)
:disabled	被禁用的	button:disabled(被禁用的按钮)
:enabled	被启用的	button:enabled(被启用的按钮)
:focus	获得焦点的	input:focus(获得焦点的输入域)
:hover	鼠标悬停时	div:hover(鼠标悬停在层上面)
:link	未访问链接	a:link(未访问的链接)
:read-only	具有只读属性	input:readonly(只读的输入域)

示例：用伪类设置六个复选框的状态。当鼠标在复选框上面悬停时，变成红色实线圆角矩形；当勾选复选框后，变成蓝色点线圆角矩形。具体效果如图 3-5 所示。

图 3-5 复选框样式

代码如下：

```
<style>
    input:hover + span {              ①
        border: 1px solid #f00;
        border-radius: 4px;           ②
        color: #f00;
    }
    input:checked + span {            ③
        border: 1px dotted #00f;
        border-radius: 4px;
    }
</style>
<div>
    <input type="checkbox" name="fav" value="01"><span>足球</span>
    <input type="checkbox" name="fav" value="02"><span>游泳</span>
    <input type="checkbox" name="fav" value="03"><span>绘画</span>
    <input type="checkbox" name="fav" value="04"><span>舞蹈</span>
    <input type="checkbox" name="fav" value="05"><span>阅读</span>
    <input type="checkbox" name="fav" value="06"><span>拳击</span>
</div>
```

说明：
① 鼠标悬停在复选框上时，改变<span>文字状态。
② 设置为圆角矩形边框。
③ 复选框被选中后，改变<span>文字状态。

2. 伪元素

伪元素用于设置页面元素指定部分的样式，例如设置一段文字的首行为红色字体，或者设置<div>层中临时动态添加的文字等。表 3-6 列出了伪元素。

表 3-6 伪元素

名　　称	说　　明	举例(作用)

::after	指定元素之后	span::after {
		　　content: url("image/m0.png");
		}
		(在<span>内部内容的末尾插入一张图片)

::before	指定元素之前	div::before { content: "欢迎光临!";}(在<div>内容前插入文字)
::first-letter	首字母	p::first-letter { color: #f00;}(段的首字母红色)
::first-line	首行	p::first-line { color: #f00;}(段的首行红色)
::selection	用户选择的内容	p::selection { color: #f00;}(用户选中的内容为红色)

示例：模拟微信的聊天框，如图 3-6 所示。

代码如下：

图 3-6 聊天框

```
<style>
    .chat_msg {                                              ①
        position: relative;
        width: 150px;
        height: 22px;
        font-size: 0.6em;
        background: #9EEA6A;
        border-radius: 4px;
        padding: 2px;
    }
    .chat_msg::after {                                       ②
        position: absolute;
        content: "";                                         ③
        top: 100%;
        left: 99%;
        margin-top: -18px;
        border: 6px solid;
        border-color: transparent transparent transparent #9EEA6A;   ④
    }
</style>
<div class="chat_msg">今天大家讨论得很热烈啊!</div>
```

说明：

① 设置绿色背景聊天框的样式。

② 定义一个伪元素——一个小方块，修饰<div>层。

③ 伪元素的内容为空字符。

④ 伪元素边框——左边框为绿色实线，其他边透明。

3.3.6 多列

多列允许将文字拆分成几栏，类似于 Word 中的分栏效果。表 3-7 列出了常用的多列属性。

表 3-7 多列属性

名称	说明	举例(作用)
column-count	列数	column-count:3;（定义为 3 栏）
column-fill	填充方式	column-fill:balance|auto;（平衡或顺序自动填充）
column-gap	列间隙	column-gap:10px;（列间隙为 10 像素）
column-rule	列规则	column-rule: 1px solid #f00;（列之间 1 像素、红色实线）
column-rule-color	列规则（颜色）	column-rule-color: #f00;（列颜色红色）
column-rule-style	列规则（样式）	column-rule-style: solid;（列样式实线）
column-rule-width	列规则（宽度）	column-rule-width: 1px;（列宽度 1 像素）
column-span	跨越的列数	column-span:all;（跨越全部列）
column-width	列宽	column-width:20px;（列宽 20 像素）

示例：将一段出师表古文，用三栏形式展示，如图 3-7 所示。

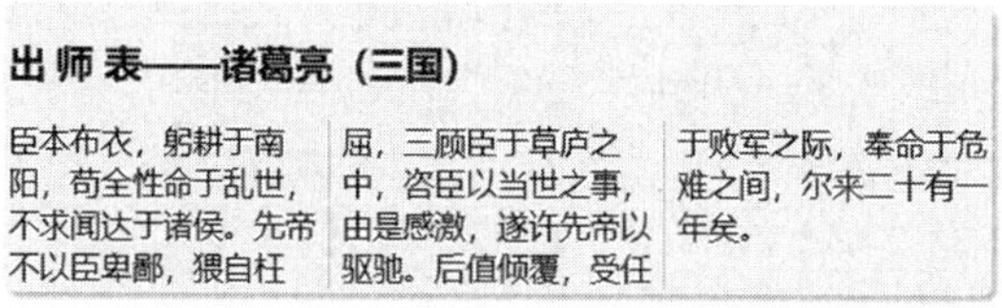

图 3-7 文字分栏

代码如下：

```
<style>
    .csb {
        margin: auto;
        width: 380px;
        font-size: 0.6em;
        column-count: 3;
        column-gap: 10px;
        column-rule: 1px solid #ccc;
        border-radius: 4px;
        background-color: #e6f3fc;
        padding: 2px;
        box-shadow: 2px 2px 2px #ccc;
    }
    h2 {
        column-span: all;
    }
</style>
<div class = "csb">
    <h2> 出 师 表——诸葛亮(三国)</h2>
    臣本布衣,躬耕于南阳,苟全性命于乱世,不求闻达于诸侯.先帝不以臣卑鄙,猥自枉屈,三顾臣于草庐之中,咨臣以当世之事,由是感激,遂许先帝以驱驰。后值倾覆,受任于败军之际,奉命于危难之间,尔来二十有一年矣。
</div>
```

3.3.7 动画

通过从一种样式变为另一种样式,CSS 也可实现一些基本动画。我们知道,动画是通过一些关键帧的连续播放来呈现的。CSS 提供了一个重要的属性@keyframes,通过它以及其他属性的组合,就可以实现动画效果。表 3-8 列出了一些常用动画属性。

表 3-8 动画属性

名 称	说 明	举例(作用)

@keyframes	动画帧	@keyframes mymove {
		from {background-color:#f00;}
		to {background-color:#00f;}
		}
		(颜色从红色变化为蓝色,也可以使用百分比)

animation-delay	延迟时间	animation-delay:2s(延迟 2s 开始)
animation-direction	播放方向	animation-direction:normal|alternate;(正常播放或反向播放)
animation-duration	持续时长	animation-duration:2s;(持续 2s)
animation-iteration-count	播放次数	animation-iteration-count:infinite;(重复播放)
animation-name	动画名称	animation-name:mymove;(动画名称 mymove)
animation-timing-function	速度曲线	animation-timing-function:linear;(恒定速度。还可取值为:ease-in 低速开始、ease-out 低速结束、ease-in-out 低速开始和结束)

示例:图片的路径动画。一张图片,放置在层<div>里面。动画轨迹:①→②→③→④→①。在运动过程中,层<div>旋转前进,且背景颜色发生变换,如图 3-8 所示。图 3-9 为动画过程中的某一帧画面。

图 3-8 动画轨迹

图 3-9 旋转动画帧

代码如下:

```
<style>
    .move {
        width: 100px;
        height: 110px;
        background: #f00;
        position: relative;
        animation-duration:5s; /*5s 完成一轮动画*/
        animation-name:moveimg; /*动画名称*/
```

```css
            animation-iteration-count: infinite; /*循环播放*/
        }
        @keyframes moveimg{ /*定义5个关键帧*/
            0% {
                transform: rotate(0deg); /*转换: 旋转 0°*/
                background: #f00;
                left: 0px;
                top: 0px;
            }
            25% {
                transform: rotate(90deg); /*转换: 旋转 90°*/
                background: #d5d502;
                left: 300px;
                top: 300px;
            }
            50% {
                transform: rotate(180deg); /*转换: 旋转 180°*/
                background: #00f;
                left: 0px;
                top: 300px;
            }
            75% {
                transform: rotate(270deg); /*转换: 旋转 270°*/
                background: #0f0;
                left: 300px;
                top: 0px;
            }
            100% {
                transform: rotate(360deg); /*还原到起点角度*/
                background: #f00;
                left: 0px;
                top: 0px;
            }
        }
    </style>
    <div class="move"><img src="image/m1.gif"></div>
```

视频讲解

3.4 场景应用示例——功能导航条

3.4.1 应用需求

京东购物网在购物页面的右边放置了一个功能导航条。综合前面所学知识，模拟制作一个水平放置的功能导航条，如图3-10所示。

图 3-10 功能导航条

具体要求：

（1）当鼠标悬停在图标按钮上面时，颜色变为红色；

（2）"消息"图标按钮，右上角用红色小圆圈显示有 6 条新消息。这里只是模拟显示，我们将在第 13 章，真正实现消息推送；

（3）单击图标按钮时，弹出显示文字的对话框，例如"查询……"。当然，这个弹出对话框功能只是示例性质，暂时没有任何实际意义，但体现了基本的处理思路，以后可以慢慢完善按钮单击功能的实际处理。

3.4.2 实现思路

首先定义一个层< div >，容纳整个工具栏图标按钮。其次，图标按钮使用< button >是最合适不过的了。最后，红色消息小圆圈，用层< div >表示，辅以 CSS 属性 border-radius 修饰，并定义好相对于父级< button >的位置即可。

3.4.3 CSS 代码

代码如下：

```
<style>
    .navi {/* 定义最外围层的样式 */
        margin: auto;
        width: 335px;
        height: 50px;
        font-size: 0.6em;
        text-align: center;
        column-count: 6;
        column-gap: 1px;
        column-rule-style: solid;
        column-rule-color: #fff;
    }
    button {/* 图标按钮的样式 */
        border: 0px;
        width: 55px;
        height: 50px;
        border-radius: 4px;
        background-color: #bfbbbb;
    }
    button:hover {/* 鼠标悬停在按钮上背景色变为红色 */
        background-color: #de2525;
    }
    .msg {/* 定义右上角的消息提示 */
        position: relative;
        left: -4.8em;
        top: 6px;
        border-radius: 50%;
        text-align: center;
```

```
            display: inline - block;
            color: #fff;
            background - color: #f15555;
            width: 1.7em;
            height: 1.6em;
            line - height: 1.6em;
            font - size: 0.6em;
        }
</style>
<div class = "navi">
    <button onclick = "alert('新增......')">
        <img src = "image/new.png" width = "22" height = "22"><br/>新增
    </button>
    <button onclick = "alert('删除......')">
        <img src = "image/delete.png" width = "22" height = "22"><br/>删除
    </button>
    <button onclick = "alert('查询......')">
        <img src = "image/search.png" width = "22" height = "22"><br/>查询
    </button>
    <button onclick = "alert('畅聊......')">
        <img src = "image/chat.png" width = "22" height = "22"><br/>畅聊
    </button>
    <button onclick = "alert('报表......')">
        <img src = "image/statistics.png" width = "22" height = "22"><br/>报表
    </button>
    <button onclick = "alert('消息......')">
        <img src = "image/info.png" width = "22" height = "22"><br/>消息
        <div class = "msg">6</div>
    </button>
</div>
```

上述代码只是在每个部分的第1句加了注释,希望读者认真对照以前所学知识,逐句研读,完全理解每句代码的含义!

3.5 场景任务挑战——导航菜单

制作学院导航菜单,如图3-11所示。"学院导航"背景为浅绿色,鼠标移动到其上时,弹出菜单项。当在菜单项中滑动鼠标时,当前菜单(例如"人工智能学院")的背景色变为灰色。

图3-11 导航菜单

JavaScript脚本语言

良好的交互性,会给访问网页的用户带来愉快的使用体验。本章主要介绍JavaScript基本语法、变量、函数、数组、对象、控制语句、JSON等内容。熟练掌握JavaScript,可为开发交互性良好的页面打下坚实的基础。

4.1 JavaScript 简介

1992年Nombas公司开发了一种与C语言相似的C-minus-minus(Cmm)语言,后改称ScriptEase,这些为JavaScript的诞生奠定了基础。

1995年美国网景公司(Netscape)发布了JavaScript 1.0,获得巨大成功,几乎成为互联网的必备组件。1996年微软公司发布JScript,成为JavaScript语言发展过程中的重要一步。

1997年,JavaScript 1.1作为一个草案提交给欧洲计算机制造商协会(ECMA),随后推出了ECMAScript全新脚本语言。从此以后,浏览器厂商们开始努力将ECMAScript作为JavaScript实现的基准。此后,ECMAScript版本不断更新。2009年12月正式发布ECMAScript 3.1,这就是著名的ES5。2015年6月发布ECMAScript 6,即ES6。ECMAScript版本2019年6月已经发展到ES10,一直在推陈出新的道路上前进。

可以将JavaScript(简称JS)看作是ECMAScript标准的实现和扩展,是一种脚本语言,并不是Java语言的另一种版本。

4.2 JavaScript 的使用

4.2.1 页面直接使用

JS代码必须放置在<script></script>标签之间,例如:

```
<script>
    alert("Hello,JavaScript!");
</script>
```

这句弹出对话框的 JS 代码,在前面第 2、3 章作为示例使用过。我们常将<script></script>及其中的 JS 代码作为整体,放置在<head></head>或<body></body>之间。

4.2.2 使用脚本文件

可以将 JS 代码按照一些规则,例如按照模块、功能、分类等,集中放在若干扩展名为.js 的文件中,例如 home.js、query.js 等,以外部文件的方式,在当前页面中引用,例如:

```
<script src = "js/home.js"></script>
<script src = "js/query.js"></script>
```

上面两行代码,引用了当前站点 js 文件夹下的 home.js、query.js 文件。这样一来,将 JS 代码从 HTML 代码中分离出来,更易于阅读,方便维护。

对于一些第三方 JS 文件,也可以通过完整的 URL 链接地址引用其 JS 脚本,例如:

```
<script src = "https://lib.baomitu.com/vue/3.1.5/vue.global.js"></script>
```

这种方式的弊端是:一旦无法与外部网络联通,将导致 JS 脚本无法使用,影响页面的正常运行。

4.3 变量和常量

4.3.1 使用 var 和 let 声明变量

变量是存储数据值的容器,其值可以发生变化。可能使用 var 或 let 声明变量。

1. 全局变量

全局(函数之外)声明的变量的有效范围是整个脚本或函数,例如:

```
var curUser = "tom", age = 20;
let passed = true;
```

这里使用 var 关键字定义了两个变量:字符串变量 curUser、数值变量 age,用 let 关键字声明了一个逻辑变量 passed。在这两句代码后的任何脚本和函数中,均可访问这三个变量。

注意:JS 是区分大小写的。

2. 局部变量

局部变量一般只能在它们被声明的函数内或者代码块内被访问。例如:

```
{
    let age = 18;
    var addr = "192.168.1.5";
}
```

上面代码中的变量 age 只能被代码块{}内的脚本访问,代码块外无法访问 age 变量的值。不过,值得注意的是,变量 addr 的值仍然可以被代码块{}外的脚本访问。

总结一下：var、let 声明的变量,在其后的脚本、函数、代码块内均可访问,但是代码块{}内用 let 声明的变量,其有效范围仅限于该代码块{},这是 var、let 有差异的地方。

4.3.2 使用 const 声明常量

const 与 let 定义变量的方式类似,但必须在声明的同时赋值,例如：

```
const age = 18;    //age 定义了对数值 18 的引用
age = 20;          //出错,不允许重新赋值
```

也就是说,const 定义的量不能重新赋值,因此被称作"常量",但并不是说其值不能改变。const 实际上创建了一个值的引用,如果 const 定义的是一个对象,尽管不能给这个对象变量重新赋值,但是却可以给该对象的某个属性重新赋值。后面会介绍 JS 对象。

4.4 基本数据类型

JS 主要有五种基本数据类型：数值、字符串、布尔型、未定义型、空。

1. 数值型 Number

用于定义各种数字的,包括整数、小数等。例如：

```
let age = 10, pi = 3.14;
```

可以用 typeof 运算符来判断变量类型,例如 alert(typeof pi)将弹框显示 number。

2. 字符串 String

字符串是用单引号或双引号括起来的,例如：

```
let loginer = "杨过";
```

如果 typeof loginer 则返回 string。

3. 布尔型 Boolean

布尔型只有两种值：true 或 false,例如：

```
let married = true;
```

同样可以用 typeof marred 判断其类型，这时返回 boolean。

4. 未定义型 Undefined

```
let tel,email = "guoy@126.com";
```

上面声明了两个变量 tel、email，用 typeof tel 判断则返回 undefined，而 typeof email 返回 string。在 JS 里面，任何声明的变量未进行初始化时，例如 tel，则该变量的默认值就是 undefined。

5. 空 Null

空是一种特别的类型，表示不存在。例如：

```
let username = null;
```

有时候，我们在某个时点并不知道用户名是什么，所以赋值为 null。当用 typeof username 判断 username 的类型时，会返回 object。这说明 null 其实是一种对象。

4.5 函数

4.5.1 使用 function 定义函数

函数是完成特定单一功能的单元。函数定义格式如下：

```
function 函数名(参数1,参数2…){
    若干代码;
}
```

参数是传递给函数的数据，可以自定义参数的个数，也可以没有参数。例如编写一个计算圆面积的函数：

```
function area(r){
    return 3.14 * Math.pow(r,2);
}
let s = area(12.6);
```

函数 area() 通过参数 r 获得需要计算的圆半径，通过 return 返回计算结果，并赋值给 s。

4.5.2 使用箭头函数

函数还有另外一种写法：箭头函数。箭头函数允许使用更简短的语法来编写函数，上面的例子用箭头函数，可改写如下：

```
const area = r => 3.14 * Math.pow(r,2);
let s = area(12.6);
```

再看另外一个示例：

```
const aScore = (x, y, z) => {
    let total = 0;
    total = x * 1.5 + y * 0.8 + Math.pow(z, 2);
    return total;
}
let w = aScore(12, 25.5, 16.1);
```

这里的 aScore 函数传递了三个参数进行计算，并返回结果。实际上，你可能已经想到，这个函数可以简写为下面形式：

```
const aScore = (x, y, z) => x * 1.5 + y * 0.8 + Math.pow(z, 2);
```

4.6 数组

数组是一种特殊类型的对象，是有序数据的集合。对于那些有内在联系的数据，数组能够为其提供批量、便利、精练的处理方式。打个比方：数组类似于一张有抽屉的桌子，桌子相当于数组名，抽屉相当于数组的元素，可以形象地用图 4-1 表示。

图 4-1 数组

图 4-1 中的数组可以这样声明：

```
let dws = [18,"ok",true,5,16, "优"];
```

用一个单一的名称 dws 存放 6 个值，其中有数值、字符串和逻辑值，并通过下标来存取数组元素。注意：数组下标是从 0 开始的。

```
dws[0] = 21;        //将数组下标为 0(也就是第 1 个位置)元素的值修改为 21
let m5 = dws[5];    //取出数组第 5 个元素的值，赋值给变量 m5
```

表 4-1 列出了数组常用的属性和方法。

表 4-1 数组常用的属性和方法

名　　称	说　　明	举例(作用)
length	长度	dws.length;(结果为 6)
push()	向尾部添加元素	dws.push("13038325832");(在 dws 最后一个元素后添加新元素："13038325832")
unshift()	向头部添加元素	dws.unshift(85);(85 成为 dws 数组的第 0 个元素)
shift()	删除首个元素	let fw＝dws.shift();(fw 值为 18，dws 值变成["ok",true,5,16, "优"])
pop()	删除最末元素	let fw＝dws.pop();(fw 值为"优"，dws 值变成[18,"ok",true,5,16])
splice()	添加(删除)元素	dws.splice(1, 0, "!");(dws 值变成[18,"!","ok",true,5,16, "优"]) dws.splice(1, 4, "!");(dws 值变成[18,"!","优"])

续表

名称	说明	举例(作用)

concat()	合并返回新数组	let adt = ["430070","027"];
		let comm = dws.concat(adt);
		(comm 的值为[18,"ok",true,5,16,"优","430070","027"])

join()	合并为字符串	let sjoin = dws.join("/");(sjoin 的值为"18/ok//true/5/16/优")
sort()	按字母顺序排序	dws.sort();(dws 值变成[16,18,5,"ok",true,"优"])
reverse()	反转元素顺序	dws.reverse();(dws 值变成["优",16,5,true,"ok",18])
forEach()	遍历数组元素	dws.forEach(e => alert(e));(用弹框循环显示 dws 的全部元素)

4.7 对象

4.7.1 对象概述

与 Java 语言类似，除了原始值，例如 true、3.14 等，JS 几乎也是"万物皆对象"。对象由属性、方法组成，一般用大括号{}括起来，例如：

```
const em = {
    name: "管理学院",
    tel: "87800490",
    url: "em.edu.cn",
    addr: "华润路 2 号",
    introEm: function () {
        alert(this.name + ": " + this.url);
    }
}
```

对象的属性由"名称:值"对组成(例如 name："管理学院")。通过"对象名.属性名"或"对象名[属性名]"，可存取(添加)属性，例如：

```
alert(em.name);              //弹框显示"管理学院"
em.tel = "59361230";         //重新设置 tel 属性值为 59361230
delete em.addr;              //删除 addr 属性
em.logo = "image/em.png";    //添加新属性 logo 并赋值
```

与此类似，对象 em 的 introEm()方法可以这样调用：

```
em.introEm();
```

也可以使用箭头函数定义对象的方法，将上面的 introEm()方法修改成：

```
introEm: () => alert(em.name + ": " + em.url)
```

4.7.2 当前对象 this

关键字 this 指当前对象,具体取值取决于 this 所处的位置。例如:

```
< button onclick = "alert(this.innerText)">
    < img src = "image/new.png" width = "22" height = "22"><br/>新增
</button>
```

这里的 this 是指这个按钮,因为 onclick 事件由 button 按钮拥有,所以 this.innerText 是指按钮的文本"新增"。再例如:

```
const student = {
    no: "2107021208",
    tel: "13025467512",
    introTxt: function (str) {
        return this.no + ": " + str;         ①
    }
}
function getWidth() {
    return this.innerWidth;                  ②
}
```

说明:

① 这里的 this,是指 student 对象,因为 introTxt 方法由 student 对象拥有,而 introTxt()方法中的 this 代表拥有者。

② 由于 getWidth()函数由页面拥有,这里的 this 是指页面拥有者,所以 this 代表打开页面的浏览器窗口对象 Window。

综上所述,实践中需要根据 this 所处的位置进行判断。

4.7.3 窗口对象 Window

Window 对象代表浏览器中打开的窗口。需要特别说明的是,Window 实际上属于 HTML DOM(Document Object Model,文档对象模型)内容范畴,但在 JS 中经常使用,所以集成到这里介绍。Window 对象有一个重要属性 window,它是对窗口对象自身的引用,我们经常直接使用这个属性进行各种窗口处理。Window 对象提供了诸如下面所列的很多属性、方法来帮助处理浏览器窗口。

◇ innerHeight:浏览器窗口内高。
◇ innerWidth:浏览器窗口内宽。
◇ open(URL,name,features,replace):打开新窗口。
◇ close():关闭当前窗口。
◇ moveTo(x,y):移动当前窗口。
◇ resizeTo(width,height):调整当前窗口大小。
◇ alert(message):显示消息警告框。

◇ confirm(message)：显示消息确认框。

◇ prompt(text,defaultText)：显示数据输入对话框。

也可以自己给窗口对象添加专门的属性或方法，例如：

```
window.myMethod = () => alert("添加自定义方法!");
```

然后，就可以在别的地方调用该自定义方法：

```
myMethod(); //或者 window.myMethod()
```

示例1：在窗口打开 login.html。

```
<script>
    function openLogin() {
        window.open("login.html","ogin","width = 600,height = 400,resizable = no,scrollbars = no")
    }
</script>
<button onclick = "openLogin()">登录</button>
```

单击"登录"按钮，将在新窗口中打开 login.html 页面，窗口名为 login，窗口宽度为 600 像素，高度为 400 像素，禁止调整大小，不显示滚动条。

示例2：制作提示确认删除的对话框。

```
<script>
    const em = {
        name: "管理学院",
        tel: "87800490",
        url: "em.edu.cn",
        addr: "华润路 2 号",
    }
    const delConfirm = () => {
        if (confirm("确认删除地址属性?"))
            delete em.addr;
    }
</script>
<button onclick = "delConfirm()">删除</button>
```

上面的代码，会弹出对话框(简称弹框)询问是否确认删除，如图 4-2 所示。一旦单击"确认"按钮，将删除 em 对象的 addr 属性。

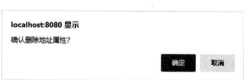

图 4-2　确认对话框

4.7.4 文档对象 Document

被浏览器载入解析的页面,都属于 document 文档对象,document 实际上是 Window 对象的属性。文档对象有几个重要方法必须掌握。

- ◇ createElement(tagName):创建指定 tagName 标签名的 HTML 元素,例如创建 HTML 图片元素 document.createElement("img")。
- ◇ getElementById(id):返回指定 id 的页面元素。
- ◇ getElementsByName(name):返回指定 name 的页面元素的集合。
- ◇ querySelector(selectors):返回与 selectors 选择器匹配的第一个 HTML 元素。
- ◇ querySelectorAll(selectors):返回与 selectors 选择器匹配的 HTML 元素列表。

注意:第二个方法返回单个对象,第三个方法返回对象集合。

示例 1:单击按钮时,将<div>中的内容修改为"你好!"。

```
<script>
    const setContent = s => {
        let div = document.getElementById("myDiv");
        div.innerText = s;
    }
</script>
<button onclick="setContent('你好!')">单击试试</button>
<div id="myDiv"></div>
```

思考:如何实现单击后将<div>中的内容修改为一张图片?这个问题看起来比较简单,但实现起来并不是简单地向 setContent()函数传送一张图片即可。请大家不妨试试!

示例 2:用模板<template>向表格中动态添加学院名称数据,页面效果如图 4-3 所示。

编号	学院名称
01	管理学院
02	会计学院
03	传媒学院

图 4-3 学院名称表格

```
<style>
    table {
        width: 180px;
        border-top: 1px solid #0099FF;
        border-collapse: collapse;
    }
    td {
        border-bottom: 1px solid #0099FF;
    }
</style>
<table>
    <tr>
        <td>编号</td><td>学院名称</td>
    </tr>
    <tbody id="slist"></tbody>
```

```
</table>
<template id="school">
    <tr>
        <td></td>
        <td></td>
    </tr>
</template>
<script>
    const schools = {
        '01': '管理学院',
        '02': '会计学院',
        '03': '传媒学院'
    }
    let template = document.querySelector('#school');
    let td = template.content.querySelectorAll("td");
    for (let k in schools) {                               //遍历 schools 对象,将其数据插入表中
        td[0].innerText = k;
        td[1].textContent = schools[k];
        let tbody = document.getElementById("slist");
        let node = document.importNode(template.content, true);   //包含子节点的深度复制
        tbody.appendChild(node);
    }
</script>
```

示例 3:单击"全选"按钮时,将复选框全部选中,如图 4-4 所示。代码如下:

☑足球 ☑电游 ☑舞蹈 ☑武术
[单击全选] [我的选择]

图 4-4 全选处理

```
<script>
    const allChecked = () => {
        let items = document.getElementsByName("item");
        items.forEach(e => e.checked = true)
    }
</script>
<div id="favs">
    <input type="checkbox" name="item" value="足球">足球
    <input type="checkbox" name="item" value="电游">电游
    <input type="checkbox" name="item" value="舞蹈">舞蹈
    <input type="checkbox" name="item" value="武术">武术
</div>
<button onclick="allChecked()">单击全选</button>
<button onclick="myChecked()">我的选择</button>
```

这里使用 getElementsByName()函数获取页面上全部 name="item"的复选框,得到一个数组 items,然后通过 forEach()循环处理 items 里面的每一个复选框对象:设置 checked 属性为 true,即勾选复选框。

思考:单击"我的选择"按钮,调用 myChecked(),显示页面上已经打对钩的复选框文本。例如:如果用户只选择了"足球""武术",则弹框显示"我的选择:足球-武术"字样,以此类推。这里并没有

给出该函数的具体实现代码。作为练习,请大家结合JS知识,查阅相关资料,挑战完成该函数。

4.7.5 事件状态 Event

敲击键盘、单击鼠标、移动鼠标……这些平时司空见惯的事件操作,由event对象来代表这些事件的状态,event同样属于DOM范畴。Event提供了较多的属性、方法,例如button、clientX、clientY、preventDefault()等,就不一一介绍了。下面,通过target属性的使用示例来了解event对象。

示例:文本的同步输出。当在"原始文本"后的文本框输入任何内容时,自动同步到"同步文本"后的文本框,实现同步输出显示效果,如图4-5所示。

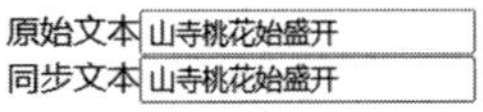

图 4-5 同步输出文本

下面是具体代码:

```
原始文本< input type = "text" id = "txt1"/><br/>         ①
同步文本< input type = "text" id = "txt2"/>
< script >
    const txt1 = document.getElementById("txt1");      ②
    txt1.onkeyup = (event) => {                         ③
        const txt2 = document.getElementById("txt2");  ④
        txt2.value = event.target.value;               ⑤
    }
</script >
```

说明:
① 定义了两个文本框,通过id属性加以区分。
② 通过id值获取第一个文本框。
③ 给第一个文本框添加键盘监听事件。当用户输入完某个字符(也即键盘onkeyup)时,调用箭头函数,将event事件对象传给该函数,以便后续处理。
④ 通过id值得到第二个文本框。
⑤ 将触发当前事件的目标元素(即第一个文本框)的内容value赋值给第二个文本框。

4.7.6 页面定位 Location

页面定位对象Location实际上是前面介绍的Window对象的属性,包含了与页面地址相关的各种处理,其用法比较简单。表4-2列出了其常用属性和方法。表4-2中的举例,基于URL地址"http://em.edu.cn:8080/schat?t=01"。

表 4-2 Location 属性方法

名称	说明	举例(作用)
host	设置(返回)主机	alert(location.host);(弹框显示"em.edu.cn:8080")
hostname	设置(返回)主机名	let host=location.hostname;(host值为em.edu.cn)
href	设置(返回)URL	location.href="login.html";(转到login.html页面)
pathname	设置(返回)路径	let path=location.pathname;(path值为"/schat")
port	设置(返回)端口号	let port=location.port;(port值为8080)

续表

名称	说明	举例(作用)
protocol	设置(返回)协议	let prc=location.protocol;(prc 值为"http:")
search	设置(返回)路径	let search=location.search;(search 值为"?t=01")
assign()	加载新文档	location.assign("index.html")(转到并打开主页 index.html)
reload()	重新加载文档	location.reload()(重新加载页面)
replace()	替换当前文档	location.replace("index.html")(用 index.html 替换当前页面)

4.7.7 样式处理 Style

Style 对象代表着样式声明，也属于 DOM 范畴。我们可以利用 Style 对象动态改变 CSS 各种属性。来看下面的示例：

```html
<script>
    const setColor = () => document.getElementById("demo").style.color = "#f00";
</script>
<div id="demo">不知细叶谁裁出</div>
<button onclick="setColor()">单击变色</button>
```

单击按钮时，通过调用 setColor() 箭头函数，将 <div> 层中的文字颜色变为红色。

4.7.8 对象包装器 Object

在 JS 中，几乎所有的对象都是 Object 的实例。下面学习几个重要方法的应用。

(1) Object.entries()：返回指定对象的键值对。

```javascript
const student = {
    no: "2107021208",
    name: "张帆",
    tel: "13025467512"
}
const listStudent = () => {
    for (const [key, value] of Object.entries(student)) {
        alert(key + "--" + value);
    }
}
```
```html
<button onclick="listStudent()">查看对象</button>
```

这里的 key 代表属性名，例如 name，而 value 则代表属性值，例如 2107021208。单击按钮后，将循环弹框显示类似于"name--2107021208"字样的内容。

(2) Object.is：判断是否为同一个值。

```javascript
let x = 12, y = "12", z = "12";
let w = x;
```

```
let r1 = Object.is(x, y);           //r1 的值为 false
let r2 = Object.is(y, z);           // r2 的值为 true
let r3 = Object.is(x, w);           // r3 的值为 true
let r4 = Object.is(null, null);     //r4 的值为 true
```

(3) Object.keys()：获得指定对象全部属性名称的集合，返回数组类型。

```
const student = {
    no: "2107021208",
    name: "张帆",
    tel: "13025467512"
}
Object.keys(student).forEach(k => student[k] = null);
```

代码获得 student 的属性名集合后，通过 forEach() 遍历数组中的每个元素，并通过属性名下标 k，将 student 对应的属性值置为 null。现在 student 对象类似于下面这样：

```
student = {
    no: null,
    name: null,
    tel: null
}
```

(4) Object.values()：获得指定对象全部属性值的集合，返回数组类型。

```
const student = {
    no: "2107021208",
    name: "张帆",
    tel: "13025467512"
}
let values = Object.values(student);
```

values 的值为 student 对象属性值的集合：["2107021208","张帆","13025467512"]。

4.8 异步操作 Promise

同步操作、异步操作是两个风格迥异的处理。打个比方，你要做两件事：看电影、吃饭。同步操作：去电影院看完电影后，才能去餐馆吃饭；或者，在餐馆吃完饭后，才能去看电影。异步操作：一边在电视上看电影，一边吃饭。

我们前面所举示例，都是同步操作。

4.8.1 Promise 对象

Promise 对象常用于异步操作的处理。Promise 有三种状态——pending：初始待定状态；

fulfilled：操作已成功完成；rejected：操作失败。Promise使用格式可以抽象化表示如下。

```
const myPromise = new Promise((参数) =>{      //执行操作A
})
    .then(B)                                  //执行操作B
    .then(C)                                  //执行操作C
    …
    .catch(出错处理);
```

这种调用被称为链式调用。当然，也可以不使用链式调用而采用常规写法。Promise提供了两个静态方法resolve(value)、reject(reason)，前者通常情况下会返回操作已成功完成状态，后者则返回一个操作失败状态。

示例：模拟现实中要执行的四个任务，其中任务2耗时较长。我们不希望任务3、任务4必须等任务2完成后才开始执行，那样太耗时间了，而是任务2开始执行后，立即执行任务3、任务4。

需要使用Promise进行异步处理，代码如下：

```
console.log("任务1--" + new Date().toLocaleTimeString());           ①
new Promise(function (resolve) {                                    ②
    setTimeout(() => resolve("任务2--" + new Date().toLocaleTimeString())
        , 2000);                                                    ③
}).then(r => console.log(r));                                       ④
console.log("任务3--" + new Date().toLocaleTimeString());           ⑤
console.log("任务4--" + new Date().toLocaleTimeString());           ⑥
```

说明：

① 执行任务1。为了方便起见，这里利用console.log()函数在浏览器控制台显示日志信息，以便观察模拟执行的情况。

② 创建Promise对象。

③ 执行任务2。通过Promise的resolve获得执行结果。利用JS的setTimeout()函数模拟耗时较长的任务2，这里模拟设置任务2耗时2s。

④ 记录任务2的执行结果。

⑤、⑥ 分别执行任务3和任务4。

运行效果需要打开浏览器控制台查看，如图4-6所示。从执行结果来看，尽管代码里面是先执行任务2，但任务3、任务4并没有等到任务2执行完后才开始执行，而是立即执行。而任务2，也在2s后执行完毕。

图4-6 异步执行

4.8.2 async和await

异步函数async允许用另一种方式构建基于Promise的异步操作，可以视需要与await搭配使用。其中，async声明该函数为异步函数，await关键字则表示阻塞并等待异步函数的执行结果。将上面的例子用async改写如下：

```javascript
const job2 = () => {
    return new Promise(resolve => {
        setTimeout(() => resolve("任务 2 -- " + new Date().toLocaleTimeString())
            , 2000);
    });
}
const doJob2 = async () => job2().then(r => console.log(r)); //使用async关键字
console.log("任务 1 -- " + new Date().toLocaleTimeString());
doJob2();
console.log("任务 3 -- " + new Date().toLocaleTimeString())
console.log("任务 4 -- " + new Date().toLocaleTimeString());
```

代码运行效果与图 4-6 效果完全一样。现在改变一下要求：任务 2 必须等任务 1 完成后才能执行，也就是变成同步执行了，如图 4-7 所示，怎么办？

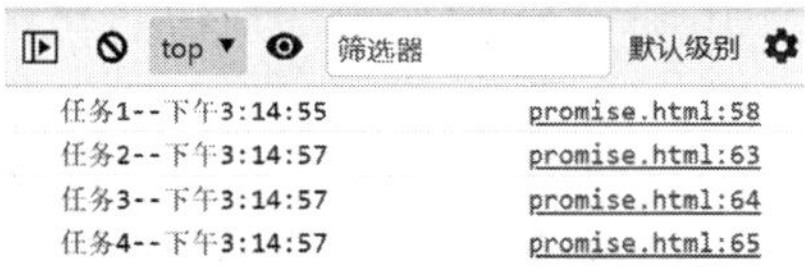

图 4-7 同步执行

使用 await 阻塞任务 2，将代码变换一下即可：

```javascript
const doJob = async () => {
    console.log("任务 1 -- " + new Date().toLocaleTimeString());
    await new Promise(resolve => {
        setTimeout(() => resolve("任务 2 -- " + new Date().toLocaleTimeString())
            , 2000);
    }).then(r => console.log(r));
    console.log("任务 3 -- " + new Date().toLocaleTimeString())
    console.log("任务 4 -- " + new Date().toLocaleTimeString());
}
doJob();
```

从图 4-7 的运行结果可知，任务 3、任务 4 等待任务 2 完成后才执行。

4.9 控制语句

4.9.1 导入（import）和导出（export）

1. import

语句 import 用于导入需要的模块。当需要加载一些 JS 代码的依赖项时，import 是最优选择。基本语法格式如下：

```javascript
import { 模块名 as 别名 } from "js 脚本文件";
```

示例：导入 js 文件中的组件。

```javascript
import {loginComponent as login, registComponent as regist} from "./users.component.js";
import {schoolsComponent as school} from "./module/schools.component.js";
```

第1句代码从当前目录(./)下的 users.component.js 脚本文件导入两个模块：loginComponent、registComponent，并分别设置简略别名 login、regist；第2句代码，则从 module 文件夹下的 schools.component.js 文件导入 schoolsComponent 模块，设置其别名为 school。

注意：import 只能在声明了 type="module" 的 <script> 标签中才能使用，例如：<script src="home.js" type="module"></script>，这样才可以在 home.js 中使用 import。

2. export

关键字 export 用于从模块中导出原始值、对象、函数等，以便其他模块在需要的时候可通过 import 导入使用。

1）单个导出

用 export 直接修饰原始值、对象或函数等，例如：

```
export const homeLoginer = {//导出对象
        username: null,
        logo: null
}
export const resetLoginer = (data) => {//导出函数
        loginer.username = data.username
        loginer.logo = data.logo
}
```

2）多个导出

可以对定义的原始值、对象或函数等一次性集中导出，例如：

```
export {homeLoginer, resetLoginer}
```

3）聚合导出

从功能管理的角度出发，为了更好地管理模块的导入导出，可以创建一个专门的 JS 文件，只用于集中管理多模块的导出。例如，创建一个 exportModules.js 文件：

```
export {homeLoginer as loginer} from "./home.js";
export {resetLoginer} from "./users/users.component.js";
export {studentQuery as query} from "./module/student.component.js";
```

exportModules.js 文件充当了聚合器功能，将分散在各个 JS 文件中的导出对象聚合在一起。当 JS 文件需要引入使用某个对象时，只需要从聚合器文件中导入即可，例如：

```
import {loginer, query} from "./exportModules.js";
```

在第13章中将会使用聚合器文件进行模块的集中管理。

4.9.2 条件判断 if...else

条件判断语句，基本格式如下：

```
if (条件成立){
    语句序列 1;
}else{
    语句序列 2;
}
```

流程很简单：条件成立,执行语句序列 1;否则,执行语句序列 2。来看下面的代码:

```
const testIf = (x) => {
    let result = 100;
    if (x > 0) {
        result -= x;
    } else {
        result += x;
    }
    return result;
}
console.log(testIf(-25));
```

4.9.3 多条件分支 switch

当并列的条件比较多时,用 switch 更为方便,格式如下:

```
switch (表达式) {
    case 常量 1:
        语句序列 1;
        break;
    case 常量 2:
        语句序列 2;
        break;
    …
    case 常量 n:
        语句序列 n;
        break;
    default:
        语句序列 n+1;
}
```

如果表达式的值等于"常量 1",则执行语句序列 1,结束;如果表达式的值等于"常量 2",则执行语句序列 2,结束……以此类推;如果都不满足,则执行"语句序列 n+1"。来看下面的示例:

```
let f = prompt("请选择(1.白菜 2.萝卜 3.茄子 4.苦瓜): ");
switch (f) {
    case '1':
    case '2':
        alert('今天白菜、萝卜价格 2.3 元/斤!');
```

```
                break;
        case '3':
                alert('今天茄子价格 2.6 元/斤!');
                break;
        case '4':
                alert('今天苦瓜价格 1.2 元/斤!');
                break;
        default:
                alert('不存在该菜品!');
}
```

在页面中的效果如图 4-8 所示。

图 4-8 switch 运行效果

4.9.4 循环操作 for

1. for

基本格式如下：

```
for ( 初始化语句;循环条件;迭代语句) {
     语句序列
}
```

可以在 for 语句内部使用 break 或 continue 进行特殊处理：break 终止循环，而 continue 则忽略后面的语句，直接进入下一轮循环。

示例 1：计算 1+2+3+…+100。

```
let sum = 0;
for (let i = 1; i <= 100; i++)
    sum += i;
alert("总和：" + sum);
```

示例 2：计算输出九九乘法表。

```
let s;
for (let i = 1; i <= 9; i++) {
    for (let j = 1; j <= i; j++) {            //j 的变化范围为 1 - i
        s = i * j;
```

```
                document.writeln(i + " * " + j + " = " + s + " ");    //末尾空格隔开
        }
        document.writeln("< br/>");                                        //换行
}
```

在浏览器中输出效果如下：

```
1 * 1 = 1
2 * 1 = 2   2 * 2 = 4
3 * 1 = 3   3 * 2 = 6    3 * 3 = 9
4 * 1 = 4   4 * 2 = 8    4 * 3 = 12   4 * 4 = 16
5 * 1 = 5   5 * 2 = 10   5 * 3 = 15   5 * 4 = 20   5 * 5 = 25
6 * 1 = 6   6 * 2 = 12   6 * 3 = 18   6 * 4 = 24   6 * 5 = 30   6 * 6 = 36
7 * 1 = 7   7 * 2 = 14   7 * 3 = 21   7 * 4 = 28   7 * 5 = 35   7 * 6 = 42   7 * 7 = 49
8 * 1 = 8   8 * 2 = 16   8 * 3 = 24   8 * 4 = 32   8 * 5 = 40   8 * 6 = 48   8 * 7 = 56   8 * 8 = 64
9 * 1 = 9   9 * 2 = 18   9 * 3 = 27   9 * 4 = 36   9 * 5 = 45   9 * 6 = 54   9 * 7 = 63   9 * 8 = 72   9 * 9 = 81
```

2．for…in

该语句遍历对象属性以便视需要进行后续处理。格式：

```
for (属性变量 in 对象)
    语句序列;
```

为了举例方便，我们先来定义一个数组：

```
const schools = [
        {cname: "管理学院", majors: ["信息管理", "工商管理", "物流管理", "电子商务",]},
        {cname: "会计学院", majors: ["会计学", "工程造价", "财务管理"]},
        {cname: "传媒学院", majors: ["动画", "广告", "摄影", "数字媒体", "新闻传播"]},
        {cname: "经济学院", majors: ["投资学", "财政学", "金融学", "国际贸易"]}
];
```

这个数组的每个元素是一个对象。每个对象的 majors 属性又是一个数组。现在，需要在网页中以下面的形式输出 schools 数组的属性值。

```
管理学院：信息管理；工商管理；物流管理；电子商务
会计学院：会计学；工程造价；财务管理
传媒学院：动画；广告；摄影；数字媒体；新闻传播
经济学院：投资学；财政学；金融学；国际贸易
```

具体实现代码如下：

```
schools.forEach(s => {                          //遍历数组的每个元素
        for (const k in s) {                    //使用 for…in,遍历对象的每个属性
            if (k == "cname")
                document.write(s[k] + ": ");    //输出学院名称
```

```
                if (k == "majors") {
                    let m = Object.values(s[k]).join("; ");    //将专业合并成字符串并以分号隔开
                    document.writeln(m);
                }
            }
            document.writeln("<br/>");                          //学院之间换行
        });
```

3. for...of

该语句与 for...in 类似,格式:

```
for (属性变量 of 对象)
    语句序列;
```

我们来将前面的示例改成 for...of 来遍历。粗体部分为修改后的代码,其他代码没做任何改变:

```
for (const s of schools) {
    for (const k in s) {
        ...
        document.writeln("<br/>");
    }
}
```

4. for await...of

对于有异步操作的处理,使用 for...of 往往得不到正确的结果,例如:

```
const getSchool = (i) => new Promise(resolve => resolve(schools[i]));    //异步获取第 i 个学院
const listSchools = async () => {
    let arr = [getSchool(0), getSchool(1), getSchool(2), getSchool(3)];  //异步获取 4 个学院
    for (const s of arr) {    ①
        for (const k in s) {
            if (k == "cname")
                document.write(s[k] + ": ");
            if (k == "majors") {
                let m = Object.values(s[k]).join("; ");
                document.writeln(m);
            }
        }
        document.writeln("<br/>");
    }
}
listSchools();
```

在浏览器中打开页面,并不能像前面那样正常输出学院、专业数据。在存在异步操作的情况下,循环已经结束,但异步可能还在执行,就容易导致输出结果异常。这时候,需要使用 for await...of,将①处的代码修改成下面这样,就能得到正确结果:

```
for await (const s of arr) {
```

4.9.5　do...while 和 while 语句

语句 do...while 循环执行语句序列,直到条件表达式为 false。do...while 中的语句序列至少执行一次。语法格式如下:

```
do {
    语句序列;
} while(条件表达式);
```

示例:计算 1＋2＋3＋…＋100。

```
let sum = 0, i = 0;
do {
    sum += ++i;
} while (i < 100);
document.writeln("总和: " + sum);
```

而 while 语句是在条件表达式为真的前提下,循环执行语句序列。语法格式如下:

```
while(条件表达式){
    语句序列;
}
```

用 while 来重新计算 1＋2＋3＋…＋100:

```
let sum = 0, i = 0;
while (i <= 100) {
    sum += i++;
}
document.writeln("总和: " + sum);
```

4.9.6　try...catch...finally 语句

异常捕获语句实际上有三种形式:try...catch、try...finally、try...catch...finally。基本语法格式如下:

```
try {
    语句序列 1;
}catch (异常) {
    语句序列 2;
}finally {
    语句序列 3;
}
```

上述异常捕获语句的流程是：试着执行语句序列 1，一旦抛出或捕获到异常，则执行语句序列 2。无论是否有异常，都执行语句序列 3。下面来看一个模拟示例：

```
try {
    console.log("执行登录……");
} catch (error) {
    console.error(error);
} finally {
    console.log("执行完毕！");
}
```

打开网页后，在浏览器控制台可以看到如图 4-9 所示的日志信息。接下来我们模拟一下出错场景，修改代码，在粗体代码下面增加一行：

```
throw "无法连接数据库！";
```

抛出异常后，控制台信息如图 4-10 所示。

图 4-9　正常执行　　　　　　　图 4-10　发生异常

从上面模拟执行过程可以看到，finally 总是会被执行，因此可以将一些无论什么情况都需要执行的代码，放在 finally 代码块里面，例如关闭数据库连接。

4.10　表单数据 FormData

在实践中，经常使用 JS 处理页面中的表单数据，因此需要学习使用 Web API（Application Programming Interface，应用编程接口）中提供的 FormData。FormData 用键值对的数据构造方式来表示表单数据。

4.10.1　通过表单< form >创建

先用 HTML 标签构建表单：

```
< form id = "login" method = "post" action = "tj.html">
    用户名< input type = "text" name = "username">
    密  码< input type = "password" name = "password">
    < button type = "button" onclick = "showFormData()">登录</button >
    < input type = "reset" value = "重置">
</form>
```

然后，在 showFormData() 函数里面构建 FormData：

```
<script>
    const showFormData = () => {
        let login = document.getElementById("login");        //获取表单元素
        const formData = new FormData(login);                //通过表单创建FormData数据
        for (const d of formData) {                          ①
            console.log(d[1]);
        }
        login.submit();                                      ②
    }
</script>
```

说明：

① 遍历表单数据，通过控制台输出用户在表单中输入的数据。实践中，可在这里对表单数据进行检查，例如必填项是否都填写完整等，当合乎要求后再提交表单。在浏览器中打开页面后，我们假定在 username 用户名文本框中输入"杨过"、password 密码框中输入"123456"，则 FormData 会构建形式为："username":"杨过"、"password":"123456"的键值对。这样一来，d[0]代表键名，例如 username，d[1]则代表键值，例如杨过。因此，单击"登录"按钮后，控制台输出数据如图 4-11 所示。

② submit()方法，用于提交表单数据给 tj.html 进行后续处理。若要在浏览器控制台观察图 4-11 的运行效果，请在本语句前面加"//"将其暂时注释掉。

图 4-11　FormData 输出

4.10.2　用代码生成 FormData

4.10.1 节创建 FormData 的方式需要跟 HTML 表单< form >挂钩，但有时候需要单独构建表单数据，以便将这些数据提交到后台进行处理，可以直接用代码生成 FormData：

```
const formData = new FormData();
formData.append("username", "admin");
formData.append("password", "123456");
formData.append("role", "teacher");
```

这样就以键值对数据形式创建了用户名、密码、用户角色三类数据，并封装成 FormData。

4.10.3　处理文件数据

文件上传在实际中应用非常普遍，那么如何用 FormData 传送文件数据？

假定有这样的场景：文件上传时，限制单文件大小为 20MB，如图 4-12 所示。如果任何一个文件超过 20MB，禁止上传，且弹框提示用户超限了；如果两个文件都没超过限制，则显示上传进度指示图片和文字。

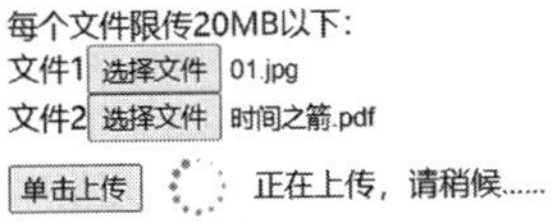

图 4-12　文件上传

HTML 页面代码比较简单，如下所示：

```html
每个文件限传20MB以下：<br/>
文件1<input type="file"><br/>
文件2<input type="file"><br/>
<div style="width:360px">
    <button onclick="fileUpload()">单击上传</button>
    <span id="progress" style="visibility:hidden">
        <img src="image/progress.gif" width="48" height="48" style="vertical-align:middle"/>正在上传,请稍候......</span>
</div>
```

页面代码里面，<span>用来显示上传进度指示图片和文字，默认是隐藏的。最关键的是fileUpload()函数，来看具体代码：

```javascript
<script>
    const fileUpload = () => {
        let num = 2;                                              //默认可正常上传的文件数
        const myfiles = document.querySelectorAll("input[type='file']");
                                                                  //全部待上传文件
        myfiles.forEach(f => {
            if (typeof f.files[0] == "undefined") {              //未选择任何文件
                num--;                                            //可正常上传的文件数减1
                return;
            }
            if (f.files[0] != "undefined" && f.files[0].size > 1024 * 1024 * 20) {
                                                                  //限传20MB
                alert("文件：" + f.files[0].name + " 超过限制大小!");
                num = 0;
                //隐藏进度指示
                document.getElementById("progress").style.visibility = "hidden";
                return;
            }
        });
        if (num > 0) {
            const formData = new FormData();
            formData.append("username", "jack");                 //添加上传用户
            myfiles.forEach(f => formData.append("myfile", f.files[0]));
                                                                  //添加上传文件
            //显示进度指示
            document.getElementById("progress").style.visibility = "visible";
        }
    }
</script>
```

代码里面的files[0]，代表每个待上传文件。这个例子还可以修改成上传更多的文件。当然，这里并没有真正向后台服务器传送文件，只是收集待上传的文本数据和文件，模拟上传而已，毕竟我们还没学习服务器端知识。

4.11 使用 JSON

4.11.1 JSON 简介

JSON(JavaScript Object Notation,JavaScript 对象表示法)是一种用于文本数据存储或交换的格式。JSON 在实践中得到非常广泛的应用,例如 A 页面临时保存用户信息、前端向后端服务器传送数据等,都可以采用 JSON。JSON 优点很多:

◇ 轻量级,比 XML 更小、更快,更易解析。
◇ 使用 JavaScript 语法来描述数据对象。
◇ 独立于语言和平台。
◇ 纯文本,没有结束标签,读写速度快。

4.11.2 JSON 基本语法

JSON 可表示数字、布尔值、字符串、null、数组以及对象,但不支持复杂数据类型。实践中,我们用 JSON 语法规则处理数据时,主要遵从以下几条规则:

◇ 数据保存在名称/值对中。
◇ 数据由逗号分隔。
◇ 对象用大括号{}表示。
◇ 数组用[]表示。

示例 1:用两个对象,以 JSON 数据格式表示两个学生。

```
{"sno":"007" , "sname":"杨过", "sclass":"电商 2101"}
{"sno":"008" , "sname":"孙大升", "sclass":"信管 2101"}
```

示例 2:格式化输出 JSON 数据。

若有下面的 wtt 对象:

```
const wtt = {
    "管理学院": [
        {"no": "007", "name": "杨过美", "class": "电商 2101"},
        {"no": "008", "name": "孙大升", "class": "信管 2101"},
        {"no": "009", "name": "张宏志", "class": "工管 2101"}
    ],
    "会计学院": [
        {"no": "001", "name": "杨过人", "class": "财管 2101"},
        {"no": "002", "name": "张曼丽", "class": "造价 2101"}
    ]
}
```

请格式化输出 wtt 对象中的数据,具体要求:

(1) 每个学院的名称作为题头,红色字体;

(2) 学生名单排列输出在各自学院名称下,淡色背景,每个学生记录用点线包围;

(3) 修饰效果也用 JS 代码实现,而非使用.css 样式文件修饰。具体效果如图 4-13 所示。

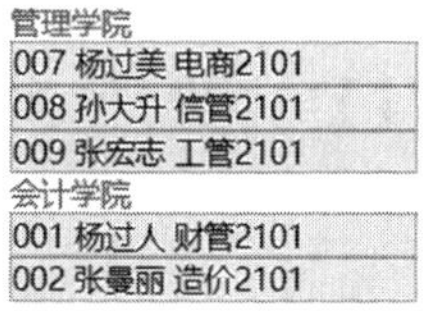

管理学院
007 杨过美 电商2101
008 孙大升 信管2101
009 张宏志 工管2101
会计学院
001 杨过人 财管2101
002 张曼丽 造价2101

图 4-13 格式化输出

由于 wtt 对象的每个 JSON 键值对代表了某个学院,因此处理思路是:遍历 wtt→获取学院→遍历学院数组→获取每个对象→输出对象的 JSON 属性值。可表示如下:

```
for (const s in wtt) {          //遍历 wtt 对象的 JSON 键值对
    ①创建一个外部<div>,用来显示题头,并用于容纳下面的内部<div>
    wtt[s].forEach(e => {       //遍历每个学院的学生记录数组
        ②创建一个内部<div>,用于容纳学生记录。该层作为外部层的子元素。
        for (const k in e) {    //遍历每条学生记录对象的 JSON 键值对
            ③创建<span>,用于显示获取到的每个学生对象的 JSON 属性值,并将该
            <span>元素添加到内部层中
        }
        ④将内部层添加到外部层中
    });
}
```

现在我们来将代码补充完整。①处的代码如下:

```
let root = document.createElement("div");
root.style.width = "160px";
root.style.color = "#f00";
root.style.fontSize = "0.6em";
root.innerText = s;
document.body.appendChild(root);
```

将创建好的<div>对象进行 CSS 修饰后,通过 document.body.appendChild(root)语句,将其作为子元素,添加到页面的<body></body>中。

② 处的代码如下:

```
let stu = document.createElement("div");
stu.style.width = "160px";
stu.style.backgroundColor = "#e8f1fb";
stu.style.border = "1px dotted #137CF1FF";
stu.style.float = "left";
stu.style.color = "#000";
```

③ 处的代码如下:

```
let span = document.createElement("span");
span.innerHTML = e[k] + " ";
stu.appendChild(span);
```

第 2 句代码稍微注意一下,因为我们在前面①处代码的第 5 行用的是 innexText。这里用 innexText 是否可行?请大家思考!

④ 处的代码只有一句:root.appendChild(stu);

大家可以将全部代码录入计算机,从整体上把握处理技术,并结合运行结果体会。

4.11.3 解析为 JSON 对象

JSON.parse()方法用于解析 JSON 字符串,返回相应的 JS 值或对象。

示例 1:JSON 字符串解析。

```
let wh = JSON.parse('"武汉"');                        //wh 的值"武汉"
let arr = JSON.parse('[10, true,"ok"]');              //arr 的值[10, true,"ok"]
const s = JSON.parse('{"name":"张三", "age":22}');    //s 的值{"name":"张三","age":22}
let x = s.age;                                        //x 的值 22
```

示例 2:解析 JSON 字符串,对其中 em 属性的属性值加 2。

```
let sch = JSON.parse(
    '[{"em": 172,"ac":180,"year":2020},\        ①
      {"em": 177,"ac":176,"year":2021}\
     ]',
    (k, v) => k === "em" ? v + 2 : v            ②
);
```

说明:

① 需要解析的字符串内容是一个数组,里面包含 2 个对象,每个对象由 2 组键值对组成。由于整个字符串内容较长,这里用反斜线"\"表示换行。

② 这是一个转换函数,目的是在转换前进行一些额外操作。其中,k 对应 JSON 数据中的属性名,v 对应属性值。如果属性名是 em,则将属性值加 2 后返回,否则原值返回。

最终,sch 是一个转换后解析到的数组对象:

```
[{"em": 174,"ac":180,"year":2020},{"em": 179,"ac":176,"year":2021}]
```

留给大家一道思考题:如何将这个数组对象的数据在页面上以下面的形式输出?

```
em: 174    ac: 180    year: 2020
em: 179    ac: 176    year: 2021
```

4.11.4 转换为 JSON 字符串

JSON.stringify()方法将 JS 值或对象转换为 JSON 字符串。

示例 1:转换为 JSON 字符串。

```
let wh = JSON.stringify("武汉");                           //wh 的值'"武汉"'
let arr = JSON.stringify([10, true,"ok"]);               //arr 的值'[10, true,"ok"]'
const s = JSON.stringify({"name":"张三", "age":22});      //s 的值'{"name":"张三","age":22}'
```

示例 2：选择性转换为 JSON 字符串。

```
let str = JSON.stringify(
    [{"em": 172, "ac": 180, "year": 2020},
     {"em": 177, "ac": 176, "year": 2021}
    ], ["em", "year"]);
```

可以使用数组,例如["em","year"],指示 JSON.stringify()转换时需要保留的属性名(其他剔除),这里只保留了 em、year 这两个属性。因此转换后字符串 str 的值为：

```
'[{"em":172,"year":2020},{"em":177,"year":2021}]'
```

也可以使用转换函数将["em","year"]替换成：

```
(k, v) => k === "ac" ? undefined : v
```

转换后字符串 str 的结果一样。

视频讲解

4.12 场景应用示例——动态增删书目

4.12.1 应用需求

某书店需要对书目进行动态增删管理。具体要求：
(1) 默认情况下,页面上有一行空白书目。该书目不能被删除；
(2) 用户可以随时单击"添加"图标,动态增加一行。添加后,光标自动定位到书名文本框,以便输入；
(3) 单击"删除"图标,删除当前行所在书目,删除前需要用户确认；
(4) 奇偶数行书目的背景色交替变化；
(5) "保存"功能暂不需要实现。打开网页后,效果如图 4-14 所示。

图 4-14 新增书目

4.12.2 处理思路

这个问题乍看起来有点复杂,其实不然。来看看处理思路:

(1) 将每一行的书目用<div>层容纳起来,作为一个整体使用,即书名文本框、单选按钮、删除图标,都是该<div>层的子元素。

(2) 由于页面上始终有一行书目,最简单的处理方法是:每次单击"添加"图标时,将最近书目所在<div>层克隆一份,添加到页面即可,这样就无须分别添加书名文本框、单选按钮、删除图标了。

(3) 单击"删除"图标删除时,实际上只要删除容纳该图标按钮所在的<div>层,即可实现整行的删除。要做到这一点,还是比较容易实现的。既然"删除"图标属于该<div>的子元素,可通过JS的this.parentNode获取<div>层,然后删除之。

(4) 由于页面上每本书都有自己的单选按钮,因此需要一个计数器变量i,给每行书目的单选按钮设置不同的name属性,否则单选按钮会出现混乱。

4.12.3 实现 HTML 页面

主要代码如下:

```
<table>
    <tr>
        <td colspan="2" class="titleTd" nowrap>
            <span style="font-size: 17px; color: #FFFFFF">新增书目</span></td>
    </tr>
    <tr>
        <td nowrap>书 名</td>
        <td id="books" nowrap>                    ①
            <div>                                 ②
                <input type="text" name="bname" autofocus>
                <input name="r0" type="radio" value="01" checked/>社会科学
                <input name="r0" type="radio" value="02"/>自然科学
                <img src="image/delete.png" title="删除书目" style="visibility: hidden"/>  
                                                  ③
            </div>
        </td>
    </tr>
    <tr>
        <td colspan=2 class="buttonTd">
            <img src="image/add.png" onclick="addBook()" title="新增书目"/>  
            <img src="image/save.png" onclick="" title="保存入库"/>
        </td>
    </tr>
</table>
<script src="addbook.js"></script>
```

说明:

① 放置书目的books单元格。后续动态添加的书目,都被添加到该单元格里面。

② 定义了一个层,用来容纳构成每行书目的书名文本框、单选按钮、删除图标。这个层的内容就是默认的第 1 行书目。

③ 第 1 行书目的删除图标被隐藏了,因为第 1 行书目不可删除。

4.12.4 编写 JS 脚本文件

下面是 addbook.js 的代码。

```javascript
let i = 1;                                              //计数器变量
const addBook = () => {
    let books = document.getElementById('books');       //放置全部书目的单元格
    let newBook = books.lastElementChild.cloneNode(true); //克隆最近行的书目及其事件
    newBook.children[0].value = '';                     //书名文本框置空
    newBook.children[1].name = 'r' + i;                 //设置社会科学单选按钮名称
    newBook.children[2].name = 'r' + i;                 //设置自然科学单选按钮名称
    newBook.children[3].style.visibility = 'visible';   //删除图标可见
    newBook.children[3].onclick = function () {         //添加删除图标的单击事件
        let delBook = this.parentNode;                  //获取上一级<div>层
        if (confirm('确认删除【' + delBook.children[0].value + '】?')) {
            delBook.previousElementSibling.children[0].focus(); //上一行书名获得焦点
            delBook.parentNode.removeChild(delBook);    //删除当前行
        }
    };
    books.appendChild(newBook);                         //新书目添加到单元格中
    newBook.children[0].focus();                        //光标定位到书名文本框
    i++;
}
```

这里并没有提供修饰页面效果的 CSS。作为练习,留给大家自行完成。

4.13 场景任务挑战——勾选删除

对书目删除任务做一点改变:使用复选框,即删除被勾选的书目,如图 4-15 所示。沿用前面思路并稍做改变:页面上必须至少有一本书,但第一本书是可以删除的。勾选删除操作,技术上需要考虑两点:

(1) 当用户全部勾选后,需要提示不能全选,至少要保留一本书目,即禁止全选删除操作;

(2) 需要获取哪些书目被勾选了,并删除之。

图 4-15 勾选删除

第 5 章 JSP基础

JSP是一种动态网页开发技术。本章主要介绍了JSP基本语法、内置对象、Servlet、应用监听器、与数据库交互等内容。熟悉JSP,有助于建立B/S模式下前后端交互处理的初步思维,为后续章节的学习做好准备。

5.1 JSP 概述

5.1.1 JSP 简介

JSP(Java Server Page,Java 服务页面)是由原美国 Sun(Sun Microsystems)公司倡导、很多公司(例如 Oracle、IBM 等)参与建立、基于 Java 语言的动态 Web 网页开发技术。

由于JSP是基于Java语言的,先天就具有多平台支持特性,在实践中仍然具有一定市场,例如中国工商银行、华夏基金等官网就是用JSP编写的。

5.1.2 JSP 基本页面结构

JSP 文件的扩展名是.jsp,基本页面结构如下:

```jsp
<%@ page contentType="text/html;charset=UTF-8" language="java" %>
<html>
<head>
    <title>Title</title>
</head>
<body>
<%
    out.print("Hello,JSP!");
%>
</body>
</html>
```

整个页面可以概要地划分成三大部分：

(1) JSP 指令部分。

第 1 行代码，属于 JSP 的页面指令。该指令一般放在第 1 行，指明了页面内容是纯文本的 HTML，字符集编码为 UTF-8，使用的语言是 Java。

(2) JSP 代码部分。

粗体部分代码，就是 JSP 代码，用于执行各种 JSP 操作。

(3) HTML 代码部分。

除了第 1 行代码、粗体部分代码，其他都属于 HTML 代码内容。当然，也可以加入前面学习过的 CSS 和 JS 等内容。

5.1.3 配置 Tomcat 依赖

为了后续示例方便，先创建好一个 Maven Webapp 项目 chapter5（若忘记创建方法，请回看第 1 章内容）。

由于我们使用 Tomcat 作为 JSP 的应用服务器，因此需要在项目中加入 Tomcat 依赖。打开项目 pom.xml 文件，输入 Tomcat 依赖代码：

```xml
<dependency>
    <groupId>org.apache.tomcat</groupId>
    <artifactId>tomcat-catalina</artifactId>
    <version>9.0.50</version>
</dependency>
```

注意添加的位置！单击图 5-1 中画圈处的 Load Maven Changes 按钮，重新加载。

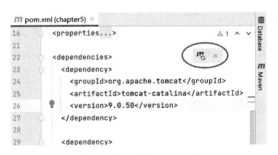

图 5-1　添加 Tomcat 依赖

5.2　JSP 基本语法

5.2.1　程序段

这里所说的程序段，是指包含变量、方法、表达式等被<%和%>括起来的部分，不包含前面 5.1.2 节中所说的 JSP 指令。如果要在项目中创建 test.jsp，代码如下：

```
<body>
<%
    int sum = 0;
    for (int i = 0; i <= 100; i++)
        sum += i;
%>
<b>计算结果:</b><%= sum %>
</body>
```

这段代码中,JSP 程序段和 HTML 代码混在一起。启动 Tomcat,在浏览器中打开页面:http://localhost:8080/chapter5/test.jsp,将在浏览器中输出"**计算结果:5050**"。

5.2.2 表达式

JSP 表达式可以将数据转换成字符串,直接输出到页面,其语法格式如下:

```
<%= 表达式 %>
```

注意:中间不能有任何空格和分号。
示例:使用表达式动态输出页面链接。

```
<body>
<%
    String[] url = new String[3];
    url[0] = "https://www.hust.edu.cn";
    url[1] = "https://www.whu.edu.cn/";
    url[2] = "index.jsp";
%>
<a href="<%= url[0] %>" target="_blank">华中科技大学</a>
<a href="<%= url[1] %>" target="_blank">武汉大学</a>
<a href="<%= url[2] %>" target="_self">首页</a>
</body>
```

5.2.3 JSP 中的注释

注释是对程序的说明性文字,为程序的可读性及日后系统的维护带来极大方便。JSP 注释不会执行,也不会发送给用户,即用户是看不到的。

```
格式 1: <%-- 注释内容 --%>
格式 2: <%//注释内容 %>
```

下面是一个完整的 JSP 页面,包含了两种格式的注释。加粗字体是格式 1 注释,有下画线的是格式 2 注释:

```jsp
<%--
    编写者：杨过
    说明：示例程序
--%>
<%@ page contentType = "text/html;charset = UTF-8" language = "java" %>
<%-- 导入日期处理相关的类 --%>
<%@ page import = "java.text.SimpleDateFormat" %>
<%@ page import = "java.util.Date" %>
<html><head>
    <title>注释示例</title>
</head>
<body>
<%-- 显示图片、格式化的日期 --%>
<%
    out.println("<img src = 'images/m0.png'>");    //显示图片
    Date today = new Date();
    //格式化
    SimpleDateFormat sdf = new SimpleDateFormat("yyyy-MM-dd HH:mm:ss");
    //输出到页面
    out.print("<h2>" + sdf.format(today) + "</h2>");
%>
</body></html>
```

通过注释，即使不熟悉相关代码的人，也可以很快理解代码的含义，提高了可读性。

5.3 JSP 内置对象

内置对象是隐含对象，无须声明和创建，可以直接使用。

5.3.1 out

out 对象主要用于向客户端发送数据，在前面其实已经接触到。out 对象提供了很多方法，这里介绍其中的 2 个常用方法。

◇ print()/println()：输出数据/输出数据后换行。
◇ close()：关闭输出流。

示例：页面停顿 3s 后返回主页。

```jsp
<body>
<%
    out.print("<span style = 'color:#f00'>请稍候，3 秒后返回到主页……</span>");
    out.print("<script>");                                                    ①
    out.print("setTimeout(\"location.href = 'index.jsp'\",3000);");           ②
    out.print("</script>");
    out.close();
    out.println("<b>欢迎再回来!</b>");                                         ③
```

```
%>
</body>
```

从上面代码可以体会到 out 对象的强大输出功能：不但可以输出普通文字，还可以输出 HTML 的各种标签，例如<span>，甚至可以输出 JS 脚本代码！

① 输出<script>脚本声明。
② 输出 JS 代码的 setTimeout()函数，停顿 3s 后，通过 location.href 将页面转向 index.jsp。
③ 这句欢迎再回来的文字，不会在页面显示，因为前面的 out.close()已经关闭输出了。

提示：尽管可以在 JSP 代码中用 out 输出 JS，但不建议这样做，这会导致代码的可读性变差，日后的维护也会极其困难。限于我们目前掌握的知识有限，目的只是演示一下 out 的强大输出能力而已。

5.3.2 request

请求对象 request 可获取通过 HTTP 传送到客户端的数据。下面是 request 的常用方法。
◇ getRequestURI()：获得所请求的 URL 地址。
◇ getServerName()：获得服务器名称。
◇ getServerPort()：获得服务器提供的 HTTP 服务的端口号。
◇ getRemoteAddr()：获得 IP 地址。
◇ setCharacterEncoding("编码类型")：设置请求的编码类型。
◇ getRemoteHost()：获得主机名，一般为 IP 地址。
◇ setAttribute(String s,Object o)：创建一个新属性并赋值为 o。
◇ getAttribute(String s)：获取指定属性名的值。
◇ getParameter(String name)：获得客户端传给服务器端的参数值。
◇ getRequestDispatcher(String var)：将请求转发到由 var 指定的页面。
◇ request.getParameterMap()：获得客户端请求数据的键值对。

示例 1：在主页 index.jsp 中打印相关信息。

```jsp
<body>
<%
    String uri = request.getRequestURI();
    String serverName = request.getServerName();
    int port = request.getServerPort();
    String ip = request.getRemoteAddr();
    String host = request.getRemoteHost();
    out.print(uri + " " + serverName + " " + port + " " + ip + " " + host);
%>
</body>
```

页面上将显示如下内容：

```
/chapter5/index.jsp localhost 8080 127.0.0.1 127.0.0.1
```

示例2：制作包含用户名、密码的表单页面login.jsp，提交给doLogin.jsp后显示用户输入的用户名、密码。

先看login.jsp的主要代码：

```
<body>
<form name="form1" method="post" action="doLogin.jsp">
    用户名<input type="text" name="usr">
    密 码<input type="password" name="pwd">
    <button>登 录</button>
</form>
</body>
```

大家对代码应该比较熟悉：通过表单，以post模式提交用户名、密码数据给doLogin.jsp。再看doLogin.jsp的代码：

```
<body>
<%
    request.setCharacterEncoding("utf-8");            //必须,否则用户名是中文时,会输出乱码
    String username = request.getParameter("usr");    //接收客户端传送的用户名
    String password = request.getParameter("pwd");    //接收客户端传送的密码
    out.print("你输入的用户名：" + username + " 密码：" + password);
%>
</body>
```

这个例子，显示了数据在不同页面间传送，以及服务端进行接收的基本方法。

5.3.3 response

响应对象response将数据从服务器端发送到客户端，以响应客户端的请求。下面是response的几个常用方法。

◇ addHeader(String name, String value)：添加指定名称和值的HTTP文件头信息。

◇ setHeader(String name, String value)：设置指定名称和值的HTTP文件头信息。

◇ sendRedirect(String url)：重定向到url地址，即打开url所代表的网页。

◇ sendError(int code)：发送错误信息编码，例如常见的"404"表示网页找不到。

图5-2 响应为Word文档

示例1：将JSP中的古诗自动下载为Word文档，效果如图5-2所示。

由于下载为Word文档，所以需要设置HTTP头，指示浏览器，将文件内容作为Word文档下载。首先，需要设置JSP响应内容为"application/msword; charset=GB18030"，其中GB18030在第1章接触过。然后，还需

要设置 HTTP 的 Content-Disposition 响应头,指示浏览器页面内容的展示形式:inline(内联形式,即作为网页的一部分)、attachment(附件形式,下载并保存到本地)。显然,我们需要设置为附件形式,并指定文件名为 poem.doc:

```
response.setHeader("Content-Disposition", "attachment;filename=poem.doc");
```

新建文件 mypoem.jsp,主要代码如下:

```
<body>
<%
    response.setContentType("application/msword;charset=GB18030");
    response.setHeader("Content-Disposition", "attachment;filename=poem.doc");
    out.print("<b>芙蓉楼送辛渐</b><br/>");
    out.print("【唐】王昌龄<br/>");
    out.print("寒雨连江夜入吴,<br/>");
    out.print("平明送客楚山孤.<br/>");
    out.print("洛阳亲友如相问,<br/>");
    out.print("一片冰心在玉壶.");
    out.close();
%>
</body>
```

在浏览器地址栏输入 http://localhost:8080/chapter5/mypoem.jsp,将自动下载为 Word 文件 poem.doc。

示例 2:根据登录用户名、密码的不同,打开不同页面。修改上 5.3.2 节中的 doLogin.jsp 文件:当用户名为 admin、密码为 007 时,打开 admin.jsp 页面,否则显示"用户名或密码错误,登录失败!"。

只需要修改 doLogin.jsp 的代码如下:

```
<%
    request.setCharacterEncoding("utf-8");
    String username = request.getParameter("usr");
    String password = request.getParameter("pwd");
    if (username.equals("admin") && password.equals("007"))
        response.sendRedirect("admin.jsp");
    else
        out.print("用户名或密码错误,登录失败!");
%>
```

5.3.4 session

会话 session 代表服务器与客户端所建立的状态信息。由于 HTTP 协议是无状态协议,即服务器并不知道客户端的状态,例如用户登录后,直接关闭了浏览器,这时候服务器并不知道。另外,购物时,当在 A 页面购买一双鞋子,再到 B 页面去购买袜子,这时候因为 HTTP 无状态性质,无法知道 A 页面购买了什么,因此需要 session 来保存客户端状态信息。每个打开网站的用户,都有自己私有的 session。下面是 session 的几个常用方法。

◇ setAttribute(String s,Object o):设置 session 属性名为指定值。
◇ getAttribute(String s):获取指定属性名的 session 值。
◇ removeAttribute(String s):删除指定属性名的 session。
◇ invalidate():使 session 失效。
◇ isNew():是否为新的 session。
◇ setMaxInactiveInterval(int time):设置 session 的有效期限,单位为秒。

示例:在主页 index.jsp 中放置三个链接:用户登录,链接到 login.jsp;用户注销,链接到 logout.jsp;文件下载,链接到 download.jsp,登录成功的用户才可以打开该链接,否则弹框提示需要先登录。

我们用 session 在不同页面之间分享信息,下面是主要代码。

1. index.jsp

```jsp
<body>
<a href="login.jsp">用户登录</a>
<a href="logout.jsp">用户注销</a>
<%
    Object loginer = session.getAttribute("loginer");        //loginer 用于记录用户登录状态
    if (loginer == null || !loginer.equals("sucess"))         //success 表示登录成功状态
        out.print("<a href='#' onclick=alert('请先登录!')>文件下载</a>");
    else
        out.print("<a href='download.jsp'>文件下载</a>");
%>
</body>
```

大家应该已经注意到,文件下载链接是根据 session 的值动态生成的。

2. login.jsp

与前面 5.3.2 节中的一样。

3. doLogin.jsp

```jsp
<%
    request.setCharacterEncoding("utf-8");
    String username = request.getParameter("usr");
    String password = request.getParameter("pwd");
    out.print("<script>");
    if (username.equals("admin") && password.equals("007")) {
        session.setAttribute("loginer", "sucess");           //用 session 设置用户登录状态
        out.print("alert('登录成功!');");
    } else
        out.print("alert('用户名或密码错误,登录失败!');");
    out.print("location.href='index.jsp'");                  //转向打开主页
    out.print("</script>");
%>
```

4. logout.jsp

```
<%
    session.invalidate();              //令 session 失效
    response.sendRedirect("index.jsp");   //转向主页
%>
```

5.3.5 application

服务器启动后,会自动创建 application 对象。与 session 属于每个用户独有不同,application 存放公共数据,网站的所有用户都可存取其数据,也就是说 application 是网站所有用户共享的。下面是 application 的三个常用方法。

◇ setAttribute(String s,Object o):设置指定名称、值的 application 对象。
◇ getAttribute(String s):获取指定属性名的 application 值。
◇ removeAttribute(String s):删除指定属性名的 application。

示例:用 application 实现网站访问量,如图 5-3 所示。

这个计数器的每位数字,用< span >显示,并用 CSS 定义了一个样式 numSpan。JSP 代码如下:

您是第 ⑥ ③ ⑤ ⑦ ④ ④ ⑥ 位访问者!

图 5-3 访问量计数器

```
<%
    if (application.getAttribute("counter") == null) {         ①
        application.setAttribute("counter", "6357438");
    }
    String counter = application.getAttribute("counter").toString();   ②
    int i = Integer.parseInt(counter);
    i++;
    application.setAttribute("counter", i);                    ③
    out.print("您是第");
    for (int j = 0; j < counter.length(); j++) {               ④
        char c = counter.charAt(j);
        out.println("< span class = 'numSpan'>" + c + "</span>");
    }
    out.print("位访问者!");
%>
```

说明:

① 服务器启动后,首次访问主页时,counter 并不存在,所以设置并赋初值。这里为了演示效果,设置了一个比较大的数。

② 获取当前 counter 的值,并转换为字符串。注意 application.getAttribute("counter")返回的是 Object 对象。

③ 访问量加 1 后,要记住这个新的值,所以给 counter 重新赋以新值。

④ 遍历总访问量字符串中的每位数字,用 CSS 修饰过的< span >显示在页面上。

样式 numSpan 代码如下：

```css
<style>
    .numSpan {
        border-radius: 50%;
        text-align: center;
        display: inline-block;
        color: #fff;
        background-color: #f15555;
        width: 1.8em;
        height: 1.7em;
        line-height: 1.7em;
        font-size: 0.6em;
    }
</style>
```

这个计数器其实有两个问题：

(1) 一旦服务器停止运行，再次启动时计数器将还原为初值；

(2) 每刷新一次页面，计数器将加1。第一个问题，可以用以后的数据库知识解决：将访问量存放到数据库里面，从数据库里面读取，这样一来服务器的启停就毫无影响了。至于第二个问题，用本章前面所学知识，即可完美解决。作为练习，请大家思考完成。

5.4 使用 Servlet

5.4.1 Servlet 简介

Servlet 是基于多线程技术、运行于服务端的 Java 类。由于 Servlet 是编译好的类，运行速度比 JSP 文件要快，因为 JSP 文件最终都要被 Tomcat 服务器编译成 Servlet 再运行的。这个动态编译的过程，是需要消耗时间的。

在浏览器中打开站点主页 index.jsp，实际上 Tomcat 是这样处理的：先将 index.jsp 的内容自动生成名为 index_jsp.java 的 Servlet 文件，然后再编译成 index_jsp.class，最终运行这个 index_jsp.class。

如果将 chapter5 项目发布到 Tomcat 的 webapps 下，启动站点后打开目录 D:\apache-tomcat-9.0.50\work\Catalina\localhost\chapter5\org\apache\jsp，就可以看到 index_jsp.java 和 index_jsp.class 这两个文件。

5.4.2 Servlet 生命周期

Servlet 从被服务器创建到销毁的过程称为生命周期。整个生命周期，如图 5-4 所示。

对读者而言，日常应用会更多地关注于 Servlet 的响应处理以及输出响应信息的过程。这会涉及 Servlet 的两个重要方法。

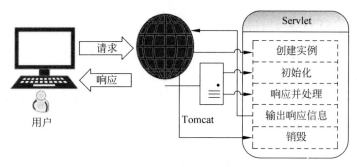

图 5-4　Servlet 生命周期

5.4.3　doGet()和 doPost()方法

doGet()方法用来处理客户端的 get 方式请求,而 doPost()自然是处理来自客户端的 post 请求。关于 get、post 这两种方式的区别,请参阅第 2 章 2.4.6 节的内容。这两个方法,都是 HttpServlet 抽象类的内部方法,需要我们覆盖这两个方法。形式如下:

```java
@Override
protected void doGet(HttpServletRequest request, HttpServletResponse response) throws ServletException, IOException {
    //在这里完成我们的各种任务
}
@Override
protected void doPost(HttpServletRequest request, HttpServletResponse response) throws ServletException, IOException {
        doGet(request, response);
}
```

读者可能注意到:在 doPost()方法里面调用了 doGet()方法!这是一种增强 Servlet 适应性的技巧。这样一来,我们的任务代码,只需要写在 doGet()方法里面。无论用户采用 get 还是 post 方式提交数据,无须修改代码,Servlet 程序都能适应。

5.4.4　加入 Servlet 依赖

要方便创建 Servlet,需要加入 Servlet 依赖。打开 pom.xml,加入以下代码并更新 Maven:

```xml
<dependency>
  <groupId>javax.servlet</groupId>
  <artifactId>javax.servlet-api</artifactId>
  <version>4.0.1</version>
  <scope>provided</scope>
</dependency>
```

5.4.5 创建 Servlet

我们需要创建一个包 servlets,用来存放 Servlet 文件。先来创建 java 文件夹,这个文件夹是必需的,以后创建其他各种包,都必须放在 java 文件夹下。创建方法:在项目的 src→main 上右击,再选择 New→Directory,在弹出的对话框中选择 java 即可,如图 5-5 所示。

图 5-5 新建 java 文件夹

现在,在 java 文件夹上右击鼠标,再选择 New→Package,完成 servlets 包的创建。

示例 1:创建名为 GoHome 的 Servlet,在页面上显示"请稍候,3s 后返回主页……"。过了 3s,自动打开主页 index.jsp。

(1) 在 servlets 包上右击鼠标,再选择 New→Servlet,如图 5-6 所示。

(2) 弹出 New Servlet 对话框。在 Name 框中输入 GoHome,再单击 OK 按钮完成创建,如图 5-7 所示。

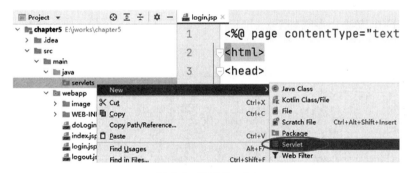

图 5-6 新建 Servlet

图 5-7 输入 Servlet 名称

(3) 完整代码如下:

```
@WebServlet(name = "GoHome", value = "/toHome")          ①
public class GoHome extends HttpServlet {                ②
    @Override
    protected void doGet(HttpServletRequest request, HttpServletResponse response) throws ServletException, IOException {
        response.setContentType("text/html;charset=utf-8");   ③
        PrintWriter out = response.getWriter();               ④
        response.setHeader("refresh", "3;URL=index.jsp");     ⑤
        out.println("<span style='color:#ff0000;font-size:0.8em'>请稍候,3s 后自动返回主页……</span>");
        out.close();
    }
    @Override
    protected void doPost(HttpServletRequest request, HttpServletResponse response) throws ServletException, IOException {
        doGet(request, response);
    }
}
```

说明:

① 这是一个注解(有些教程称为注释),用来标注该类是一个 Servlet,同时设置 GoHome 的映射名称为 toHome。映射,就是将 Servlet 虚拟成 Web 网络访问名,这意味着 GoHome 类的 Web 访问地址是:http://localhost:8080/chapter5/toHome。可以将映射名设置成需要的形式,例如,如果将 value = "/toHome"修改成 value = "/to/go.html",则该 Servlet 的访问地址变成:http://localhost:8080/chapter5/to/go.html。实际上,go.html 在站点中并不存在,这就是所谓的"映射"。后面,我们还会用到各种注解。

② 所有的 Servlet 类,包括 GoHome,都是 HttpServlet 抽象类的子类。

③ 设置响应内容为纯文本 HTML,使用 utf-8 编码。

④ 这其实就是我们前面在 JSP 文件中使用的 out 对象的真正来源。

⑤ 5.3.3 节用过 setHeader()方法,这里通过设置 HTTP 响应头的 refresh,达到 3s 后自动刷新并打开 index.jsp 的效果。

示例 2:修改前面 5.3.4 节的用户登录处理,用 Servlet 类 DoLogin(映射名 login.do)来完成登录处理。

(1) 修改 login.jsp,将原 action="doLogin.jsp"修改为 action="login.do";

(2) 新建 Servlet 类 DoLogin.java,代码如下:

```java
@WebServlet(name = "GoHome", value = "/login.do")
public class DoLogin extends HttpServlet {
    @Override
    protected void doGet ( HttpServletRequest request, HttpServletResponse response ) throws ServletException, IOException {
        response.setContentType("text/html;charset=utf-8");
        request.setCharacterEncoding("utf-8");
        PrintWriter out = response.getWriter();
        String username = request.getParameter("usr");
        …
        request.getSession().setAttribute("loginer", "sucess");
        …
    }
}
```

大部分代码与 5.3.4 节 doLogin.jsp 中的代码相同,予以省略。加粗代码是 Servlet 中设置 session 属性值的方式,与 doLogin.jsp 中是有差异的,提请注意。

5.5 EL 表达式语言

5.5.1 EL 概述

EL(Expression Language,表达式语言)的目的是使 JSP 更简单、更易读。EL 的使用,需要 JSTL(JSP Standard Tag Library,JSP 标准标签库)的支持。

EL 的使用方式非常简单:

```
${表达式}
```

来看看 JSP、EL 在页面输出"你好,EL!"的差别。JSP 是这样的:

```
<body><% out.print("你好,EL!"); %></body>
```

而 EL 则是这样:

```
<body>${"你好,EL!"}</body>
```

5.5.2 加入 JSTL 依赖

在项目 pom.xml 中加入 JSTL 依赖:

```
<dependency>
    <groupId>javax.servlet.jsp.jstl</groupId>
    <artifactId>javax.servlet.jsp.jstl-api</artifactId>
    <version>1.2.2</version>
</dependency>
<dependency>
    <groupId>org.apache.taglibs</groupId>
    <artifactId>taglibs-standard-impl</artifactId>
    <version>1.2.5</version>
</dependency>
```

5.5.3 内置对象

EL 提供了一些可以直接使用的内置对象,下面列出了一些常用的 EL 内置对象。
◇ param:客户端请求参数对象。
◇ pageScope:当前页面属性对象。
◇ requestScope:本次请求属性对象。
◇ sessionScope:会话对象。
◇ applicationScope:应用对象。
后面结合 EL 的条件输出或循环输出,再举例说明对象的使用方法。

5.5.4 条件输出

根据条件是否成立来决定内容的显示与否。
单条件表达式:

```
<c:if test="表达式">
    ...
</c:if>
```

多条件分支表达式：

```
<c:choose>
    <c:when test="表达式">
        …
    </c:when>
    ……
    <c:otherwise>
        …
    </c:otherwise>
</c:choose>
```

这里面的表达式，判断的是服务端的某个变量、对象、session 等，下面结合示例来学习其用法。

示例：用 EL 表达式改写 5.4 节中的登录示例。

首先，改写 5.4.5 节中的 DoLogin 类，去掉 Servlet 中输出 JS 代码的不合理做法；然后，修改主页 index.jsp：用户登录操作完毕后，用红色文字显示登录成功与否的提示。

1) 修改 DoLogin 类的 doGet()方法

```
@Override
protected void doGet (HttpServletRequest request, HttpServletResponse response) throws
ServletException, IOException {
    request.setCharacterEncoding("utf-8");
    String username = request.getParameter("usr");
    String password = request.getParameter("pwd");
    if (username.equals("admin") && password.equals("007"))//登录成功
        request.getSession().setAttribute("loginer", "sucess");
    else                                                    //登录失败
        request.getSession().setAttribute("loginer", "fail");
    response.sendRedirect("index.jsp");                     //转向打开主页
}
```

现在代码简洁了很多。在服务端用 session 属性 loginer 记住用户登录状态：若登录成功，其值设置为 sucess；登录失败则设置为 fail。在前端，JSP 页面用 EL 表达式，取出 session 的值，并进行判断。

2) 修改主页 index.jsp

为了方便与 5.3.4 节中的代码进行比较，下面给出了完整代码：

```
<%@ page contentType="text/html;charset=UTF-8" language="java" %>
<%@ page isELIgnored="false" %>                                        ①
<%@ taglib prefix="c" uri="http://java.sun.com/jsp/jstl/core" %>
<html>
<head><title>主页</title></head>
<body>
<a href="login.jsp">用户登录</a>
<a href="logout.jsp">用户注销</a>
```

```
<c:choose>                                                                    ②
    <c:when test = "${sessionScope.loginer == 'sucess'}">                     ③
        <a href = 'download.jsp'>文件下载</a>
        <span style = "color:#FF0000"><br/>登录成功!</span>
    </c:when>
    <c:when test = "${sessionScope.loginer!= 'sucess'}">                      ④
        <a href = '#' onclick = alert('请先登录!')>文件下载</a>
        <c:if test = "${sessionScope.loginer == 'fail'}">
            <span style = "color:#FF0000"><br/>用户名或密码错误,登录失败!</span>
        </c:if>
    </c:when>
</c:choose>
</body></html>
```

说明:

① 这句页面指令,启用 EL 表达式。紧跟的下面这句指令,则定义使用 JSTL 标签库。要在 JSP 页面使用 EL,不要忘了放置这两条加粗字体的指令!

② 使用多条件分支表达式,进行登录成功与否判断,也可以改为<c:if>语句。

③ 利用 EL 内置对象中的 sessionScope,获取后台 Servlet 类 DoLogin 中设置好的 loginer 值,进行判断。这里也可简写为:<c:when test = "${loginer == 'sucess'}">。

④ loginer! = 'sucess',包含两种情况:用户压根就没进行登录操作,这时候 loginer 为 null 值,不能在页面显示登录失败的类似字样;用户进行了登录操作,但失败,这时候 loginer 值为 fail。

图 5-8 登录成功

图 5-8 是登录成功后的效果图。使用 EL 表达式后,index.jsp 页面不再有任何 JSP 代码块,可读性更好。请大家与 5.3.4 节中的 index.jsp 进行比较,充分理解代码的含义。

5.5.5 循环输出

循环输出语法格式如下:

```
<c:forEach var = "变量" [items = "集合"][begin = "起始位置"][end = "结束位置"] [step = "步长"]>
    ...
</c:forEach>
```

中括号[]中的内容是可选的。下面的示例,动态输出 5 行文本框:

```
<c:forEach var = "i" begin = "1" end = "10" step = "2">
    文本框${i}<input type = "text" name = "txt${i}" placeholder = "EL${i}…"><br/>
</c:forEach>
```

文本框1	EL1......
文本框3	EL3......
文本框5	EL5......
文本框7	EL7......
文本框9	EL9......

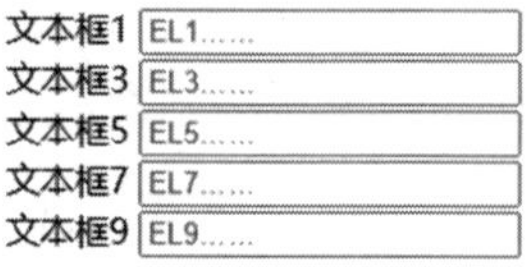

图 5-9 动态生成文本框

在网页中的效果如图 5-9 所示。

5.6 监听器

5.6.1 监听器类型

监听器(Listener)是一种特殊的Servlet类,专门用来监听Web应用或特定Java对象的初始化或销毁、方法调用、属性改变等事件。当被监听者发生相应事件后,监听器会立即执行事先实现的方法。主要有以下三种类型的监听器。

◇ ServletContextListener:Web应用监听器。主要有两个方法——contextInitialized(),Web应用发布时调用此方法,可在此方法中做一些初始化工作;contextDestroyed(),Web应用被注销或服务器停止时被调用,可在此方法中做一些扫尾工作。在第7章会使用这个监听器。

◇ HttpSessionListener:会话session监听器。主要有两个方法——sessionCreated(),创建session调用;sessionDestroyed(),销毁session时调用。

◇ HttpSessionAttributeListener:session属性监听器。主要有三个方法——attributeAdded(),向session中添加新属性时调用;attributeRemoved(),从session中移除属性时被调用;attributeReplaced(),session属性被替换时调用。

5.6.2 基于监听器的在线用户统计

用监听器统计在线用户,基本思想是:先定义一个HashMap对象,存放到application对象里面。当有用户打开站点时,sessionCreated()被触发,就可以在sessionCreated()里面将代表该用户的session id存入HashMap对象里面。若用户超时导致session失效,会触发sessionDestroyed(),可在此将对应的session id从HashMap对象中移除。这样,HashMap对象的容量大小,就表示在线用户的数量,而JSP页面通过application.getAttribute()即可获取到该数量值。

在servlet包上面右击鼠标,在弹出的快捷菜单中选择New→Web Listener,输入监听器类名称AppListener,如图5-10所示。

修改AppListener类,主要代码如下:

图5-10 创建监听器

```
@WebListener //注解为监听器
public class AppListener implements ServletContextListener, HttpSessionListener {
                                                                                  ①
    private Map<String, HttpSession> map = new ConcurrentHashMap(1);               ②

    @Override
    public void contextInitialized(ServletContextEvent sce) {
        sce.getServletContext().setAttribute("online", map);                        ③
    }
    @Override
```

```
        public void contextDestroyed(ServletContextEvent sce) {
            map.clear();                              //清除内容
        }
        @Override
        public void sessionCreated(HttpSessionEvent se) {
            HttpSession session = se.getSession();  获取 session 对象
            session.setMaxInactiveInterval(60 * 20);                                    ④
            map.put(session.getId(), session);        //当前 session 存入 map
        }
        @Override
        public void sessionDestroyed(HttpSessionEvent se) {
            map.remove(se.getSession().getId());      //移除 session
        }
    }
```

说明:

① 只需要使用两个监听器类:Web 应用监听器、会话 session 监听器。

② 新建一个线程安全、支持高并发的 ConcurrentHashMap 对象,以便保存 session。这里将 HashMap 对象初始值设为 1,后面往 HashMap 里面 put 内容时,其容量会自动增长。

③ 设置 application 对象属性 online,这样前端 JSP 页面就可通过 EL 内置对象 applicationScope 获取到 online 的值。

④ 设置 session 的有效期限为 20min。在前面学习过 HTTP 是无状态协议,只好通过 session 记录状态信息。有些网站在用户登录后,基于安全考虑,会设置了一个有效时间期限,超限则予以自动注销,就是基于设置 session 有效期限的基本原理。

现在,只需要在 JSP 页面简单使用下面的语句,即可显示在线用户数。

```
在线用户数: ${applicationScope.online.size()}
```

提示:为了保证在线用户数的准确性,建议大家将站点发布到 Tomcat 下运行,而不是在 IntelliJ IDEA 里面运行测试。

5.7 与数据库交互

数据库是计算机技术领域最重要的发展产物,是大多数信息管理系统的基础框架,是现代应用最广泛的技术之一。很多网站的背后,都有数据库技术的支持。

5.7.1 创建 users 表并加入数据库依赖

1. 创建 users 表

先设计一个用户表 users,以便后续使用,其结构如图 5-11 所示。这些字段分别表示用户名、密码、用户头像、角色(例如 teacher、student)、email 地址。

图 5-11 users 表结构

2. 加入数据库依赖

要与数据库交互,需要在 pom.xml 中加入以下数据库依赖:

```xml
<dependency>
    <groupId>org.postgresql</groupId>
    <artifactId>postgresql</artifactId>
    <version>42.2.23</version>
</dependency>
```

5.7.2 数据库连接

1. 连接原理

按照第 1 章确定的技术环境,连接数据库的过程如图 5-12 所示。

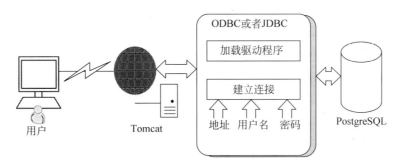

图 5-12 数据库连接原理

从图可知,ODBC 或 JDBC 驱动相当于 Web 应用与数据库之间的"桥梁"。由于 ODBC(Open Database Connectivity,开放数据库连接)连接数据库的效率低于 JDBC,且影响了跨平台特性,现实中使用较少,这里就不做介绍了。JDBC(Java DataBase Connectivity,Java 数据库连接)是由原 Sun 公司提供的统一的标准应用程序编程接口,能够与各种数据库进行交互。

2. 驱动程序

JDBC 驱动程序通常是一个或多个 jar 文件，它实际上是封装数据库访问的类库。驱动程序名一般采用"包名.类名"的形式。例如：

- ◇ org.postgresql.Driver：PostgreSQL 数据库。其中，org.postgresql 是包名，而 Driver 才是真正的驱动程序类。
- ◇ com.mysql.jdbc.Driver：MySQL 数据库。
- ◇ com.microsoft.jdbc.sqlserver.SQLServerDriver：SQL Server 数据库。

驱动程序需要进行如下加载后才能使用。

```
Class.forName("org.postgresql.Driver");
```

3. 连接地址

建立连接还需提供数据库地址 URL、能够访问数据库的用户（含用户名、密码）。URL 一般由三部分组成，各部分间用冒号分隔，格式如下：

```
jdbc:<子协议>:<子名称>
```

<子协议>是指数据库连接机制名。不同数据库厂商的数据库连接机制名是不同的，例如 MySQL 数据库使用的是 mysql，PostgreSQL 数据库使用的是 postgresql。<子名称>提供定位数据库的更详细信息，包括服务器、端口、数据库名等。例如：

- ◇ jdbc:postgresql://localhost:5432/tamsdb：通过 5432 端口连接本地服务器上的 PostgreSQL 数据库 tamsdb。
- ◇ jdbc:mysql://localhost:3306/mydb：通过 3306 端口连接本地服务器上的 MySQL 数据库 mydb。
- ◇ jdbc:microsoft:sqlserver://localhost:1433;DatabaseName=jspdb：通过 1433 端口连接本地服务器上的 SQL Server 数据库 jspdb。

5.7.3 JDBC 应用

1. 驱动管理器 DriverManager

DriverManager 是 JDBC 的管理层，作用于用户和驱动程序之间，负责加载驱动程序，建立数据库和驱动程序之间的连接。DriverManager 有三种使用格式：

- ◇ DriverManager.getConnection(String url)
- ◇ DriverManager.getConnection(String url, String user, String password)
- ◇ DriverManager.getConnection(String url, Properties info)

示例：

```
Class.forName("org.postgresql.Driver");
String url = "jdbc:postgresql://localhost:5432/tamsdb";
Connection conn = null;
conn = DriverManager.getConnection(url, "admin", "007");
```

或者：

```
Properties prop = new Properties();
prop.put("user", "admin");
prop.put("password", "007");
conn = DriverManager.getConnection(url, prop);
```

2. 连接 Connection

一个连接对象代表与特定数据库的会话。Connection 的重要方法如下。

◇ close()：关闭连接。

◇ createStatement()：创建会话声明对象。

◇ prepareStatement(String sql)：预编译 SQL 语句。

◇ setAutoCommit(boolean autoCommit)：设置 Connection 是否处于自动提交模式，默认为自动提交。如果是自动提交，则所有的 SQL 语句将被立即执行。否则，SQL 语句整体作为一个数据库的事务，由 commit() 提交，由 rollback() 撤销。

◇ commit()：提交事务。

◇ rollback()：回滚事务。撤销当前事务处理所做的任何改变，回到最初的状态。

3. 会话 Statement 和 PreparedStatement

Statement 代表发送 SQL 语句的一次会话，常用方法有以下几种。

◇ close()：关闭会话。

◇ executeQuery(String sql)：执行 SQL 语句并返回结果集。

◇ setFetchSize(int rows)：设置从数据库预读取 rows 条记录。

◇ setMaxRows(int max)：设置从数据库返回结果中结果集的最大行号数为 max。

◇ executeUpdate(String sql)：执行数据更新(update、delete、insert)SQL 语句。

◇ addBatch(String sql)：将 SQL 语句加入批量更新列表。批量更新可以提高程序执行效率，优化数据处理过程。

◇ executeBatch()：执行批量更新。

PreparedStatement 是 Statement 的子类，可存储一条预编译的 SQL 语句，并可高效、多次执行该语句。PreparedStatement 在一些场景应用中比 Statement 执行效率更高。实践中，对于重复性发生的业务，使用预编译有性能上的优势，例如用户登录，这是一种操作没变只是操作的数据发生变化的业务，建议使用预编译方法。

使用 Statement：

```
conn = DriverManager.getConnection(url, "admin", "007");
Statement s = conn.createStatement();
```

第 2 句代码可以换成预编译方式：

```
String sql = " select username,logo,role from users";
PreparedStatement ps = conn.prepareStatement(sql);
```

4. 结果集

通过 SQL 语句查询数据库后，获得需要的数据结果（结果集，或称记录集），然后再根据业务要求进行各种处理后，将数据返回给前端 JSP 页面进行显示。获取结果集 ResultSet 的方法如下：

```java
Statement s = conn.createStatement();
ResultSet rs = s.executeQuery("select username,logo,role from users");
```

或者：

```java
String sql = "select username,logo,role from users";
PreparedStatement ps = conn.prepareStatement(sql);
ResultSet rs = ps.executeQuery();
```

拿到结果集后，就可以对其中的每个对象（对应于数据表中的每条记录）进行操作。ResultSet 提供了许多 getXXX() 方法，来获取当前行对象的属性值。这里的 XXX，表示要根据属性的类型，选择相应的方法。例如：

```java
String sname = rs.getString(1);        //获取 sname 姓名
String sex = rs.getString("sex");      //获取性别
int score = rs.getInt(3);              //获取 score 分数,也可以用 rs.getInt("score")
```

一般来说，使用索引方式（获取姓名代码）而非名称（获取性别代码）执行速度更快。与此相对应，ResultSet 提供了许多 setXXX() 方法来设置属性值。除了 getXXX()、setXXX() 方法外，表 5-1 列出了 ResultSet 其他常用方法。

表 5-1 ResultSet 常用方法

方　　法	功　能　说　明
boolean next()	移动到下一行并判断是否到末尾
boolean absolute(int row)	移动到结果集指定行
void close()	关闭结果集
void deleteRow()	删除当前行
void first()	移动到结果集的第 1 行
int getRow()	获得当前行的行号
void insertRow()	插入新行
void updateRow()	更新当前行

示例 1：循环遍历 users 表的查询结果集并输出。

```java
String sql = "select username,password from users";
PreparedStatement ps = conn.prepareStatement(sql);
ResultSet rs = ps.executeQuery();
while (rs.next()) {                                             //循环遍历
    out.print(rs.getString(1) + " " + rs.getString(2));         //输出用户名、密码
}
rs.close();                                                     //关闭结果集
ps.close();
```

示例 2：修改用户 aaa 的密码为 123456。

```
String sql = "update users set password = '123456' where username = 'aaa'";
PreparedStatement ps = myconn.prepareStatement(sql);
ps.executeUpdate();
```

示例 3：修改 5.5.4 节 DoLogin 类的 doGet()方法，实现基于数据库的用户登录。
只需要修改 doGet()方法的代码如下：

```
protected void doGet(HttpServletRequest request, HttpServletResponse response) throws IOException {
    request.setCharacterEncoding("utf-8");
    String username = request.getParameter("usr");
    String password = request.getParameter("pwd");
    request.getSession().setAttribute("loginer", "fail");                       ①
    String url = "jdbc:postgresql://localhost:5432/tamsdb";
    Connection conn = null;
    try {
        Class.forName("org.postgresql.Driver");               //加载 JDBC 驱动程序
        conn = DriverManager.getConnection(url, "admin", "007");   //建立连接
        String sql = "select * from users where username = ? and password = ?";  ②
        PreparedStatement ps = conn.prepareStatement(sql);
        ps.setString(1, username);                                              ③
        ps.setString(2, password);
        ResultSet rs = ps.executeQuery();
        if (rs.next())                                       //查找到记录,登录成功
            request.getSession().setAttribute("loginer", "sucess");
        rs.close();
        ps.close();
    } catch (Exception e) { e.printStackTrace();
    } finally {
        if (conn != null) {
            try { conn.close();
            } catch (SQLException e) { e.printStackTrace();}
        }
        response.sendRedirect("index.jsp");
    }
}
```

说明：

① 用 session 记录登录状态，默认假定用户登录失败。

② 定义查询的 SQL 语句。这里的问号？是占位符，不妨这样理解：当不清楚用户名、密码的具体值时，用问号占位，等待后续传值过来。

③ 将前面接收的用户名，传值给第 1 个问号占位符。

5.8 场景应用示例

5.8.1 文件上传

文件上传是 Web 应用中很常见的功能,一般采用第三方上传组件来实现。读者也可以自力更生实现文件上传。

1. 上传数据的编码格式

浏览器在上传文件时,是按照一定数据编码格式进行的:传送数据时在头尾部分附加了额外信息,具体如下:

```
---------------------------7d429871607fe
Content-Disposition: form-data;name="表单控件名";filename="上传文件的路径和文件名"
Content-Type: text/plain
上传文件的内容……
---------------------------7d429871607fe
Content-Disposition: form-data; name="filename" Content-Type: application/octet-stream
---------------------------7d429871607fe--
```

在传送数据的头部、尾部都辅以分割线,并附加了 Content-Disposition 信息,例如 form-data、表单文件控件名、上传文件名等。如果自己编写文件上传处理,将数据写入服务器时,要忽略掉这些附加数据。

2. 编写上传 JSP 文件 upload.jsp

```html
<Script language="javascript">
    function check() {
        if(document.uploadForm.myfile.value == "") {
            alert("请选择待上传文件!");
            return false;
        }
        return true;
    }
</Script>
<form name="uploadForm" method="post" action="upload.do"
    enctype="multipart/form-data" onsubmit="return check()">
    请选择待上传文件<input type="file" id="myfile" name="myfile"><br>
    <input value="开始上传" type="submit">
</form>
```

这个表单与常规纯文本处理的表单稍有不同,请注意粗体部分的代码,用于定义文件传输的 HTTP 信息。另外,<input>的类型是 file。

3. 编写 Servlet 类 FileUpload

```java
@WebServlet(name = "FileUpload", value = "/upload.do")      //映射为 upload.do
@MultipartConfig                                             //配置上传文件支持
public class FileUpload extends HttpServlet {
    @Override
    protected void doPost(HttpServletRequest request, HttpServletResponse response) throws
ServletException, IOException {
        request.setCharacterEncoding("utf-8");
        response.setContentType("text/html;charset=UTF-8");
        String destPath = request.getServletContext().getRealPath("/upfiles");         ①
        Part part = request.getPart("myfile");            // myfile 对应表单的文件输入框
        String header = part.getHeader("content-disposition");
        String fileName = getFileName(header);            //获得文件名
        part.write(destPath + File.separator + fileName); //写入服务器
        response.getWriter().print("文件上传成功!");
    }
    private String getFileName(String header) {           //获取文件名
        String fieldName = null;
        if (header != null && header.startsWith("form-data")) {
            int start = header.indexOf("filename=\"");                                  ②
            int end = header.indexOf('"', start + 10);                                  ③
            if (start != -1 && end != -1) {
                fieldName = header.substring(start + 10, end);
            }
        }
        return fieldName;
    }
}
```

说明：

① 上传的文件都存放到站点 upfiles 文件夹下，事先需要在 webapp 下创建好该文件夹。

② 找到头部附加信息中"filename="的位置，目的是从 filename 值中获取到文件名。

③ 从"filename="后面一个字符开始（因为"filename="长度为 9），找到 filename 属性值字符串，即文件名。

5.8.2 在页面中显示 Excel 表格

使用 Servlet 显示上一节上传的 Excel 表格文件。图 5-13 为上传的 Excel 文件 reimb.xls，图 5-14 则是 Servlet 输出到网页后的效果。二者样式上会有些许差异，但整体还算不错。

1. 加入 POI 依赖

对 Excel 的处理，需要用到 Apache 软件基金会的著名 Excel 处理工具 POI。POI 提供了最完整、最正确的 Excel 格式读取实现，并采用了内存优化方式处理 Excel 表格，可以帮助读者轻松读写 Excel 文件。POI 免费、开源，可到其官网下载。在 pom.xml 中添加依赖项：

jQuery+Vue.js+Spring Boot贯穿式项目实战（微课视频版）

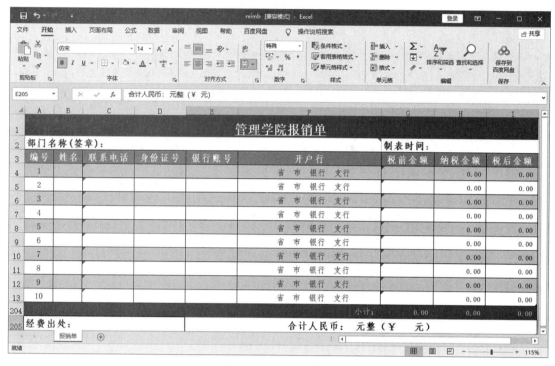

图 5-13　Excel 报销单

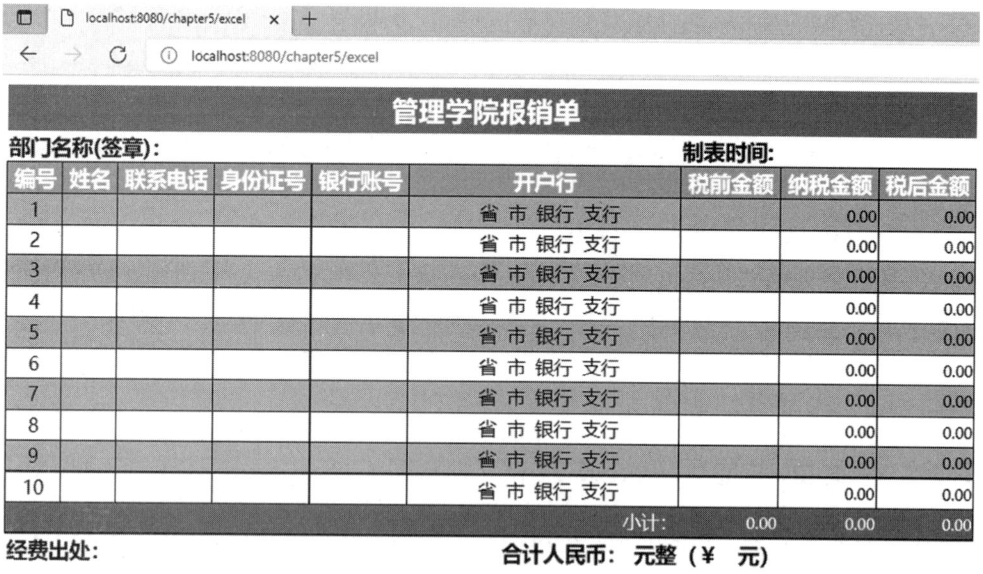

图 5-14　网页显示 Excel 报销单

```
<dependency>
    <groupId>org.apache.poi</groupId>
    <artifactId>poi</artifactId>
    <version>5.0.0</version>
```

```xml
</dependency>
<dependency>
    <groupId>org.apache.poi</groupId>
    <artifactId>poi-scratchpad</artifactId>
    <version>5.0.0</version>
</dependency>
```

2. 编写 Servlet 处理 Excel

```java
@WebServlet(name = "WebExcel", value = "/excel")            //地址映射为 excel
public class WebExcel extends HttpServlet {

    @Override
    protected void doGet(HttpServletRequest request, HttpServletResponse response)
throws IOException {
        String excelFile = request.getServletContext().getRealPath("upfiles/reimb.xls");
        HSSFWorkbook wb = ExcelToHtmlUtils.loadXls(new File(excelFile));
        wb.setSheetName(0, " ");                            //去掉 Excel Sheet 名称
        writeToHTML(response, wb);
        wb.close();
    }

    private void writeToHTML(HttpServletResponse response, HSSFWorkbook wb) {
        response.setContentType("text/html;charset=UTF-8");
        try {
            Document doc = XMLHelper.newDocumentBuilder().newDocument();
            ExcelToHtmlConverter converter = new ExcelToHtmlConverter(doc);
            converter.setOutputColumnHeaders(false);        //不显示列头
            converter.setOutputRowNumbers(false);           //不显示行号
            converter.processWorkbook(wb);

            Properties prop = new Properties();
            prop.setProperty("encoding", "UTF-8");          //设置编码
            prop.setProperty("indent", "yes");              //缩进
            prop.setProperty("method", "html");             //转换模式: html
            Transformer tf = TransformerFactory.newInstance().newTransformer();
            tf.setOutputProperties(prop);
            tf.transform(new DOMSource(converter.getDocument()),
                    new StreamResult(response.getOutputStream()));//数据流输出
        } catch (Exception e) {
            e.printStackTrace();
        }
    }
}
```

这里只是结合 Servlet,用一个小示例对 POI 抛砖引玉。实际上,POI 还可以帮助创建、读取或输出 Excel 文件,具有强大的 Excel 处理能力,读者需要在实践中不断探索、掌握其用法。

注意：上述方法只能处理 xls 文件，对于 xlsx 则需要更复杂的方法。

5.8.3 用 PDF 显示古诗

1. 应用需求

在前面已经领略到了 Servlet 强大的处理能力，接下来拓展其应用、加深理解：用 Servlet 在页面以 PDF 方式显示一首古诗。PDF 文件的应用非常广泛，很多网站常常使用 PDF 向用户展示内容。若要方便处理 PDF 文档，需要 iText 配合使用。

2. iText 简介

iText 是面向 Java、.NET 的强大的 PDF 处理工具，提供了一套通用、可编程和企业级 PDF 解决方案，并可嵌入到其他各类应用中，例如与 Servlet 结合。

iText 官网提供了开源免费版和商业版两种形式。

3. 加入 iText 依赖

要使用 iText，需先在 pom.xml 中加入依赖：

```xml
<dependency>
    <groupId>com.itextpdf</groupId>
    <artifactId>itext7-core</artifactId>
    <version>7.1.16</version>
    <type>pom</type>
</dependency>
```

4. 中文问题

iText 对中文支持不尽如人意。尽管官方也提供了处理中文的方法，但是略显遗憾的是，少数情况下某些中文会出现乱码。一个可行的解决方法是：使用 Windows 自带的字体来支持中文显示。只需要以下两个步骤。

（1）在项目 webapp 下新建文件夹 font，用于存放字体文件；

（2）将 C:\Windows\Fonts 文件夹下的某个字体文件，例如华文彩云（STCAIYUN.TTF），复制到 font 文件夹下。

下面，就可以用这个字体文件，以 PDF 文件格式显示一首古诗。

5. 具体实现

在 servlets 包下新建 Servlet 类 PdfPoem.java，主要代码如下：

```java
@WebServlet(name = "PdfPoem", value = "/pdf")          //地址映射为 pdf
public class PdfPoem extends HttpServlet {
    @Override
    protected void doGet(HttpServletRequest request, HttpServletResponse response)
            throws IOException {
        ServletOutputStream stream = response.getOutputStream();
        PdfWriter writer = new PdfWriter(stream);
        PdfDocument pdfDoc = new PdfDocument(writer);   //创建 PDF 文档对象
```

```java
        Document doc = new Document(pdfDoc, PageSize.A4);        //页面 A4 大小
        //获取字体文件的绝对路径,getRealPath()返回绝对路径
        String font = request.getServletContext().getRealPath("/font/STCAIYUN.TTF");
        PdfFont fontChinese = PdfFontFactory.createFont(font,
                        PdfEncodings.IDENTITY_H);                //创建 PDF 中文字体对象
        Paragraph header = new Paragraph("秋夕\n【唐】杜牧");    //Pdf 段落,\n 为换行
        header.setFont(fontChinese);                             //使用华文彩云字体
        header.setFontSize(16);
        header.setTextAlignment(TextAlignment.CENTER);
        doc.add(header);
        List poem = new List().setListSymbol("\u2022");          //无序列表
        poem.setFont(fontChinese);
        poem.setFontSize(14);
        poem.setTextAlignment(TextAlignment.CENTER);
        poem.add(new ListItem("银烛秋光冷画屏,"));                //每句诗作为列表项添加到列表
        poem.add(new ListItem("轻罗小扇扑流萤."));
        poem.add(new ListItem("天阶夜色凉如水,"));
        poem.add(new ListItem("卧看牵牛织女星."));
        doc.add(poem);
        doc.close();
    }
    @Override
    protected void doPost(HttpServletRequest request, HttpServletResponse response) throws IOException {
        doGet(request, response);
    }
}
```

从上述处理过程来看,iText 的代码还是非常简洁、清晰、易读的。在浏览器地址打开 http://localhost:8080/chapter5/pdf,将显示如图 5-15 所示的 PDF 文件。

图 5-15　PDF 古诗(1)

注意:这里在导入类的时候,若有多个选择,一般请选择属于 iText 的类。不少人出错,很大程度上是因为导入了错误的类。

5.9 场景任务挑战——有背景图的 PDF 古诗

在数据库中创建存放古诗的 poems 表，请合理设计出该表的相应字段。新建 Servlet 类 LatestPoem，地址映射为/poem，每次读取 poems 表中最新录入的那首古诗，然后用 6 行 1 列的无边框表格，显示整首诗。整个表格的背景需要设置成一张图片，并将字体改用其他字体，例如华文隶书，如图 5-16 所示。

图 5-16　PDF 古诗（2）

第 6 章 MVC 设计模式

设计模式(Design Pattern)是程序设计实践经验的总结和提升。设计模式使得 Web 应用可读性好,流程控制、业务逻辑和数据显示分离。本章主要介绍 MVC 设计模式的基本思想、实现过程,并结合实例应用 MVC 设计模式。熟悉并理解 MVC,有助于读者逐步建立模块化开发思想,为后续章节的学习打下较好的设计与实现基础。

6.1 MVC 概述

6.1.1 传统 JSP 开发模式

传统的 JSP 开发模式,可用图 6-1 表示。

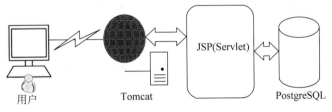

图 6-1 传统 JSP 模式

这种模式下,JSP 既负责完成业务逻辑的处理、流程的控制,还负责数据的显示。各种代码混杂在一起,很难剥离,导致后期维护或扩展比较困难。

6.1.2 MVC 原理

20 世纪 70 年代,IBM 公司就进行了 MVC 设计模式的研究,现在已经成为 Web 应用程序开发中广泛应用的软件设计模式。MVC 提供了清晰的设计框架,是模型-视图-控制器(Model View Controller)的缩写。

1. 模型（Model）

模型主要包括业务逻辑处理和数据库访问操作，例如用户登录就是一个业务逻辑处理，向数据库 users 表（对应 Java Web 中的实体类）添加一条记录就属于数据库访问操作。在 Java Web 中，一般将实体类称为 Bean 类、数据访问类称为 DAO（Data Access Object）类。

2. 视图（View）

视图是与用户交互的界面，不应该包括任何业务逻辑的处理。视图一般由 HTML、JSP、JS、CSS 等内容构成，可接收用户输入的数据，例如用户名、密码，或展示模型层返回的数据，例如登录成功或失败的标识。

3. 控制器（Controller）

控制器主要用来实现 Web 应用中各种业务流程的控制、任务的分派等，例如接收用户登录请求、调用后端的模型完成登录处理，并返回模型数据。控制器在视图层与模型层之间起到了桥梁的作用。

4. 工作流程

MVC 设计模式的实践操作：用 Servlet 作为控制器，JSP（HTML）页面作为视图层，而实体类及业务逻辑处理类作为模型层，如图 6-2 所示。

从图中可以看出，用户页面请求（V）提交给控制器（C）。控制器根据请求要求，调用模型（M）的业务逻辑处理类（实体类配合）进行处理，并获取到处理结果。最后，控制器将模型返回的结果，转发给视图（V）。在这个过程中，三者各取所长，紧密配合，完成整个业务处理过程。

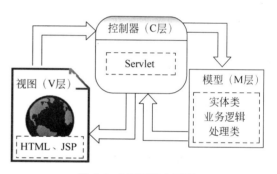

图 6-2　MVC 基本原理

6.1.3　MVC 的优缺点

作为广泛应用的设计模式，MVC 自然有其优势：

（1）低耦合性、易于维护。由于三者分离，当业务逻辑发生改变时，一般只需要改动模型层，并不涉及视图层、控制器层。程序员可集中精力于业务逻辑处理，而界面人员则集中精力于视图层上。

（2）规范化管理方便、权限控制好处理。例如，可以根据用户权限的不同，设计不同的模型层。

（3）集中管理、代码复用。例如，应用请求都由控制器层统一调度，用户模型既可以应用于用户登录视图，也可复用于用户注册视图。视图可以用 HTML、JSP 或 PDF，视图层改变了，但控制层和模型层无须改变。

当然，MVC 也有其不足之处。对于业务处理逻辑非常简单的需求，严格遵循 MVC 模式反而增加了架构的复杂性。对于业务处理逻辑复杂的需求，视图可能需要控制器与模型的多次调用，才能获取到满足要求的数据。

6.2 MVC实现过程

多数情况下,可将MVC实现过程概括为以下几步。

(1) 定义Bean来表示数据。

例如定义一个Bean类Users,表示用户数据,对应数据库中的用户表。

(2) 编写Servlet来处理请求。

例如第5章中的Servlet类DoLogin就是用来处理登录请求的。

(3) 填写Bean。

可将业务处理后得到的数据,保存到Bean中,以便后续转发给视图。

(4) 将Bean存储到请求。

在5.3.2节学过request对象的一些方法,其中request.setAttribute()方法可将数据Bean存储到请求中。

(5) 将请求转发到JSP页面。

还是利用5.3.2节学过的request对象,其中的getRequestDispatcher()方法能够将请求转发到指定页面。

(6) 从Bean中提取数据。

JSP视图可利用EL表达式,从Bean中提取数据并显示在页面上。

6.3 场景应用示例——用户注册

视频讲解

6.3.1 应用需求

某高校拟开发教务辅助管理系统,为后台管理员用户提供注册功能,如图6-3所示。

图6-3 用户注册

具体要求:

(1) 用户名唯一,不允许重复;

(2) 注册成功后打开后台管理页面admin.jsp(该页面普通用户无法直接打开);

(3)提交注册后,若检测到用户名已被注册过,则返回注册页面并记住用户所填写的信息(密码除外,被清除掉)。

此处及后续章节的一些功能处理的设计,是为了练习所学知识,对于实践中本身有些复杂的处理会进行简化设计。

6.3.2 处理思路

整个过程的基本架构如图6-4所示。

技术处理思路:

(1)用户提交注册后,先用isExisted(String username)方法检测该用户名是否已被注册过;

(2)若是未被注册的用户,调用addUser(Users user)方法将当前用户写入数据表users,并将数据写入Bean类users,转发并打开admin.jsp;

(3)若已被注册过,则仍然将用户数据填写到Bean类Users,并将密码值修改为"该用户名已被注册!",以便注册视图能够显示提示信息;

(4)注册视图使用EL表达式,绑定Bean类users中的相关信息;

(5)视图admin.jsp存放到站点WEB-INF\views文件夹下,因为WEB-INF文件夹下的内容是受到服务器保护的,无法直接打开。

现在,来创建Maven Webapp项目chapter6,添加依赖:tomcat、servlet、jstl、taglibs和postgresql驱动(跟第5章类似)。创建Java包controller、model,并准备相关图片,最终chapter6项目结构如图6-5所示。

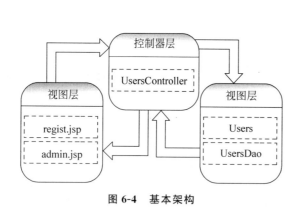

图6-4 基本架构

图6-5 项目结构

6.3.3 模型层

1. Bean类Users.java

```
public class Users implements Serializable {
    private String username;                        //用户名
```

```java
    private String password;                    //密码
    private String email;                       //邮箱
    private String logo;                        //用户头像,暂未使用,后续章节会用到
    private String role;                        //用户角色

    public String getUsername() {               //getter 方法
        return username;
    }
    public void setUsername(String username) {  //setter 方法
        this.username = username;
    }
    …//其他属性的 getter、setter 方法
}
```

这些获取(设置)类属性的方法,被称为 getter(setter)方法。这些方法并不需要手工输入,可利用右键快捷菜单 Generate→Getter and Setter,由 IntelliJ IDEA 编辑器快速生成。

2. Dao 类 UsersDao.java

先来看看 UsersDao 类的基本结构:

```java
public class UsersDao {
    private String driver = "org.postgresql.Driver";
    private String url = "jdbc:postgresql://localhost:5432/tamsdb";
    private String dbUsername = "admin";
    private String dbPassword = "007";
    //判断用户名是否已存在
    public boolean isExisted(String username) {
        boolean existed = false;
        …
        return existed;
    }
    //添加新用户
    public Users addUser(Users user) {
        …
        return user;
    }
}
```

该类首先定义了一些连接数据库有关的类变量,以便类方法使用。isExisted()方法判断是否已被注册过,addUser()方法则返回一个用户对象。

1) isExisted()方法代码

```java
public boolean isExisted(String username) {
    boolean existed = false;                    //默认用户名未被注册过
    Connection conn = null;
    try {
        Class.forName(driver);
```

```java
            conn = DriverManager.getConnection(url, dbUsername, dbPassword);
            String sql = "select username from users where username = ?";
            PreparedStatement ps = conn.prepareStatement(sql);
            ps.setString(1, username);
            ResultSet rs = ps.executeQuery();
            if (rs.next())                    //用户名已经存在
                existed = true;
            rs.close();
            ps.close();
        } catch (Exception e) {
            e.printStackTrace();
        } finally {
            if (conn != null) {
                try {
                    conn.close();
                } catch (SQLException e) {
                    e.printStackTrace();
                }
            }
        }
        return existed;
    }
```

2）addUser()方法代码

```java
public Users addUser(Users user) {
    Connection conn = null;
    try {
        Class.forName(driver);
        conn = DriverManager.getConnection(url, dbUsername, dbPassword);
        String sql = "insert into users(username,password,role,email) values(?,?,?,?)";
        PreparedStatement ps = conn.prepareStatement(sql);
        ps.setString(1, user.getUsername());
        ps.setString(2, user.getPassword());
        ps.setString(3, user.getRole());
        ps.setString(4, user.getEmail());
        ps.executeUpdate();         //执行更新,写入数据库
        ps.close();
    } catch (Exception e) {
        user.setPassword("出错,注册失败!");
    } finally {
        if (conn != null) {
            try {
                conn.close();
            } catch (SQLException e) {
                e.printStackTrace();
            }
        }
```

```
        }
        return user;
}
```

读者可能注意到,这两个方法中有些重复代码,例如连接数据库、关闭数据库。作为练习,读者可以创建一个单独的类,该类的作用很单一,即专门负责连接数据库、关闭数据库,这样就可以实现代码复用了。

6.3.4 控制器层

控制器层代码就比较简单了。在 controller 包下新建 UsersController 类,映射为"/regist.do",其 get()方法代码如下:

```
request.setCharacterEncoding("utf-8");
String username = request.getParameter("username");      //接收视图传来的用户名
String password = request.getParameter("password");
String email = request.getParameter("email");
String role = request.getParameter("role");
String goUrl = "regist.jsp";                              //控制器默认转发的页面

Users user = new Users();                                 //创建 Users 对象来表示数据
user.setUsername(username);
user.setPassword(password);
user.setEmail(email);
user.setRole(role);
if (new UsersDao().isExisted(username))                   //已被注册过
    user.setPassword("该用户名已被注册!");
else
    user = new UsersDao().addUser(user);                  //写入数据库
if (user.getPassword().equals(password))                  //密码未被重置,则设置转发页面为后台管理页面
    goUrl = "WEB-INF/views/admin.jsp";

RequestDispatcher rd = request.getRequestDispatcher(goUrl);//创建请求转发器对象
request.setAttribute("user", user);                        //将 Bean 类 user 对象存储到请求中
rd.forward(request, response);                             //转发到 goUrl 指定的页面(含数据 bean)
```

6.3.5 视图层

1. regist.jsp

```
<form name="form1" class="user-regist" method="post"
      onsubmit="return check()" action="regist.do">
    注册教务后台管理通行证<br>
    <img src="image/username.png">
    <input type="text" class="rinput" placeholder="用户名"
```

```
                    name = "username" required value = "${user.username}">
        <br><img src = "image/password.png">
        <input type = "password" class = "rinput" placeholder = "密   码"
                    name = "password" id = "pwd" required>
        <br><img src = "image/repwd.png">
        <input type = "password" class = "rinput" placeholder = "确认密码" id = "rpwd" required>
        <br><img src = "image/email.png">
        <input type = "email" class = "rinput" placeholder = "email"
                    name = "email" value = "${user.email}">
        <br><input type = "radio" name = "role" value = "student"
                        <c:if test = "${user.role == 'student'}">checked</c:if> required>学生
        <input type = "radio" name = "role" value = "teacher"
                        <c:if test = "${user.role == 'teacher'}">checked</c:if> required>教师<br>
        <button class = "rbutton">提交注册</button><br>
        <span>${user.password}</span>
</form>
```

请注意粗体部分代码,这个与后台 Bean 数据进行了绑定。对于用户角色的单选,使用了 EL 条件表达式进行判断。CSS 样式代码部分,请读者自行补充。

2. JS 函数 check()

```
<script>
    const check = () => {
        let pwd = document.getElementById("pwd").value;
        let rpwd = document.getElementById("rpwd").value;
        if (pwd != rpwd) {
            alert("两次输入的密码不一致!");
            return false;
        }
        return true;
    }
</script>
```

check()函数是在表单的 onsubmit 里面调用的。只有 return true,表单才会被提交。

3. admin.jsp

这个视图只是为了配合示例的演示,代码很简单:

```
<%@ page contentType = "text/html;charset = UTF-8" language = "java" %>
<%@ page isELIgnored = "false" %>
<%@ taglib prefix = "c" uri = "http://java.sun.com/jsp/jstl/core" %>
<html><head>
    <title>后台管理</title>
</head><body>
欢迎进入后台管理,【${user.username}】管理员!
</body></html>
```

6.4 场景任务挑战——学生信息查询

查询数据是实践中很常见的一种应用场景。请先在数据库中创建 student 表。这个表在后续章节也会使用。字段含义分别为：学号、姓名、班级、电话、地址、邮编，如图 6-6 所示。

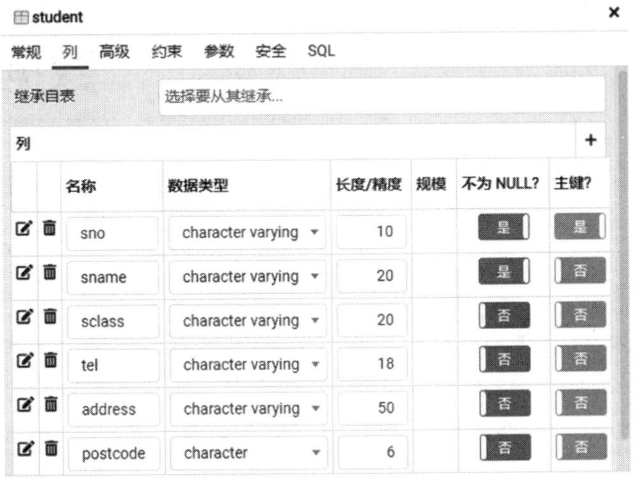

图 6-6 student 表结构

请基于 MVC 模式，实现学生信息查询：
（1）输入学生姓名，例如张三丰，若未查到数据，则显示"查无【张三丰】此人！"；
（2）若查到数据，则显示全部同名学生；
（3）查询页面，不使用 JSP 代码块，而使用 EL 表达式。运行效果类似于图 6-7。

图 6-7 查询界面

第 7 章

数据库连接池

连接数据库是 Web 应用系统中非常频繁的操作,提高数据库连接的效率可以大幅提升 Web 应用系统的性能。本章主要介绍连接池基本原理、常见数据库连接池产品,重点介绍 HikariCP 连接池及其应用。理解并掌握连接池的使用,有助于建立更为高效的 Web 应用系统。

7.1 连接池概述

7.1.1 连接池基本原理

我们在前面都是通过 DriverManager.getConnection()方法获得与数据库的连接,这个连接是一个物理连接。"物理"的含义就是在物理网络(例如 TCP/IP 网络)上建立一个用于通信的连接。显然,每次去获取物理连接并不合适。一方面,物理连接数据库是非常消耗时间的;另外一方面,多数情况下 Java Web 应用程序经常需要频繁连接数据库,无法复用物理连接。这导致常规连接数据库的方法效率低下、难以令人满意。于是,连接池技术应运而生。

可以事先建立一个任何用户都可共享连接的"连接池",每一个物理连接成为"池"中的对象(称之为池化)。当需要连接数据库时,并不需要直接向数据库发出请求,只需要到连接池中去获取一个连接对象就可以了。如果不再需要该连接对象了,则归还到池中,以供其他请求再次使用。为此,有一个称为连接池管理器的组件,专门负责管理池中的连接对象,例如连接对象的建立、管理和释放等工作。

图 7-1 说明了连接池的基本原理:

连接池管理器负责管理连接池,要管理的任务比较多:连接池创建后,初始的连接数是多少,最多可以有多少个活动(正在被使用)连接,当前未被使用的空闲连接数,空闲连接经过多长时间仍然未被使用则被销毁等。毕竟,连接池是要消耗服务器内存的。太多或太少的连接数,都可能带来问题,这往往需要根据预估访问量确定参数,并根据实践中的表现进一步调整。

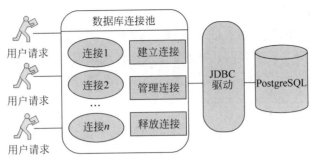

图 7-1 连接池原理

7.1.2 常见连接池产品

下面简要介绍市场上比较知名的连接池产品。

1. C3P0

C3P0 为开源、成熟、易于使用和扩展的连接池产品，支持 jdbc4 规范。著名的 ORM 产品 Hibernate 提供了 hibernate-c3p0 默认连接池支持，截至书稿完稿时最新成熟版本为 5.5.5。

2. DBCP2

DBCP2 为著名的连接池产品 DBCP1 的升级版，但与 DBCP1 并不兼容。该连接池来自著名的 Apache 软件基金会。DBCP 产品最早发布于 2002 年 8 月，以简易、高效著称，截至书稿完稿时最新版本为 2.9.0。

3. Tomcat JDBC Pool

Tomcat JDBC Pool 为 Tomcat 自带的连接池产品，性能强悍，使用方便。

7.1.3 Tomcat 连接池示例

1. 所需的类

Tomcat 连接池使用简单，主要需要如下两个类。

- org.apache.tomcat.jdbc.pool.DataSource：用于创建数据源。通过数据源，就可获取池化连接。
- org.apache.tomcat.jdbc.pool.PoolProperties：用于设置连接池的一些参数，例如驱动程序名、数据库 URL、初始活动连接大小等。

2. 加入依赖

在 pom.xml 中加入依赖：

```
<dependency>
    <groupId>org.apache.tomcat</groupId>
    <artifactId>tomcat-jdbc</artifactId>
    <version>9.0.50</version>
</dependency>
```

3. 创建 DBFactory 类

应该创建一个单独的工厂类 DBFactory。通过这个工厂类，建立连接池（只做一次），并提供 getConnection()方法，以便其他类可以源源不断地从连接池里面获取"产品"（池化的连接），而不是每次只要去访问数据库，就创建连接池并获取连接。也就是说，只有"一座"工厂，这个工厂只有"一个"连接池。下面是工厂类 DBFactory.java 的代码：

```java
public class DBFactory {
    private static DBFactory instance;            //工厂类实例对象
    private static DataSource dataSource;         //数据源

    private DBFactory() {                         //只有一座"工厂",禁止外部创建工厂类对象
        super();
    }
    public static DBFactory get() {               //获取唯一的工厂类实例对象
        if (instance == null) {
            instance = new DBFactory();
            setupPool();                          //建立连接池
        }
        return instance;
    }
    public Connection getConnection() {           //获取池化的连接
        Connection conn = null;
        try {
            conn = dataSource.getConnection();
        } catch (SQLException e) { }
        return conn;
    }
    public void closeConnection(Connection conn) {  //将连接归还到"池"中
        try {
            if (conn != null)
                conn.close();
        } catch (SQLException e) {
            e.printStackTrace();
        }
    }
    private static void setupPool() {             //构建连接池
        PoolProperties p = new PoolProperties();
        p.setDriverClassName("org.postgresql.Driver");
        p.setUrl("jdbc:postgresql://localhost:5432/tamsdb");
        p.setUsername("admin");
        p.setPassword("007");
        p.setMaxActive(15);                       //最大活动连接数
        p.setInitialSize(5);                      //初始连接数
        p.setMaxIdle(5);                          //最多空闲连接数
        p.setMinIdle(3);                          //最少空闲连接数
        dataSource = new DataSource();
        dataSource.setPoolProperties(p);
    }
}
```

4. 修改 Dao 类

现在,第 6 章中凡是连接数据库的语句,都可以将原有的:

```
conn = DriverManager.getConnection(url, dbUsername, dbPassword);
```

修改成更简洁的:

```
conn = DBFactory.get().getConnection();
```

而这个 DBFactory 类,完全可以被别人拿去复用:只需要调整 setupJdbcPool()中的参数即可。读者如果感兴趣,可以修改第 6 章的示例,改用 Tomcat 连接池。

7.2 HikariCP 连接池

7.2.1 HikariCP 简介

HikariCP 是一个"零开销"的产品级 JDBC 连接池,高性能、高品质、简单而又可靠。HikariCP 极小,只有大约 130KB,非常轻量。HikariCP 已经成为 SpringBoot 的默认连接池,在市场中的关注度非常高。本书后续示例都将使用 HikariCP 连接池。

7.2.2 加入 HikariCP 依赖

在 pom.xml 中加入以下代码:

```xml
<dependency>
    <groupId>com.zaxxer</groupId>
    <artifactId>HikariCP</artifactId>
    <version>4.0.3</version>
</dependency>
```

7.2.3 配置 HikariCP 连接池

结合前面的 DBFactory 类,对第 6 章的场景应用示例进行改造,用 HikariCP 连接池来实现数据库连接。

1. 修改 DBFactory 类

```java
public class DBFactory {
    private static DBFactory instance;
    private static HikariDataSource dataSource;
    …
    private static void setupPool() {
        Properties p = new Properties();
```

```
        p.setProperty("dataSourceClassName", "org.postgresql.ds.PGSimpleDataSource");
        p.setProperty("dataSource.user", "admin");
        p.setProperty("dataSource.password", "007");
        p.setProperty("dataSource.databaseName", "tamsdb");
        p.setProperty("dataSource.serverName", "localhost");
        p.setProperty("dataSource.portNumber", "5432");
        p.setProperty("maximumPoolSize", "20");    //最多连接数,含活动、空闲连接
        p.setProperty("minimumIdle", "5");          //最少空闲连接数
        HikariConfig config = new HikariConfig(p);
        dataSource = new HikariDataSource(config);
    }
}
```

没有变化的代码,直接省略了。数据源现在改成了 HikariDataSource,参数配置上也有些差异。

2. 修改 UsersDao 类

```
public class UsersDao {
    public boolean isExisted(String username) {
        boolean existed = false;
        Connection conn = null;
        try {
            conn = DBFactory.get().getConnection();
            String sql = "select username from users where username = ?";
            ...
        } finally {
            DBFactory.get().closeConnection(conn);
        }
        return existed;
    }
    public Users addUser(Users user) {
        Connection conn = null;
        try {
            conn = DBFactory.get().getConnection();
            String sql = "insert into users(username,password,role,email) values(?,?,?,?)";
            ...
        } finally {
            DBFactory.get().closeConnection(conn);
        }
        return user;
    }
}
```

代码简洁了很多。读者可以测试一下用户注册运行情况,一切正常。

7.2.4 查看 HikariCP 活动情况

打开 PostgreSQL 自带的数据库管理工具 pgAdmin 4,选择 tamsdb 数据库,在仪表盘可以看到连接池的连接信息,如图 7-2 所示。连接池中,当前有 5 个空闲连接。

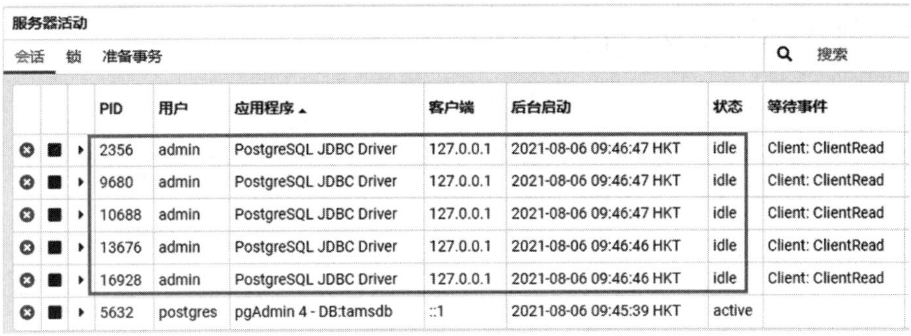

图 7-2　连接情况

也可以打开 Windows 的命令提示符，输入命令：netstat -ano|findstr 5432，查看连接池中的连接情况，如图 7-3 所示。

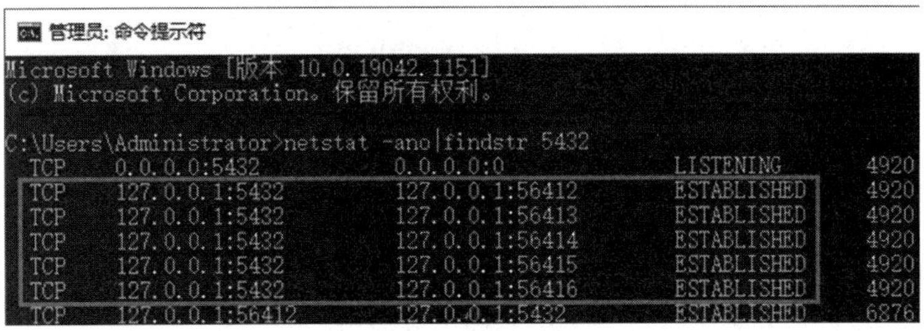

图 7-3　查看连接情况

7.3　场景应用示例——优化 HikariCP 使用

视频讲解

7.3.1　应用需求

连接池是一次建立，反复使用。那么，这个"一次建立"放置在什么时候较好？在服务器发布 Web 应用时建立，显然是最佳方案。具体来说，就是利用 Web 应用监听器 ServletContextListener：
◇ Web 应用初始化时执行如下代码。

```
contextInitialized(ServletContextEvent sce){
    //构建连接池
}
```

◇ Web 应用关闭时执行如下代码。

```
contextDestroyed(ServletContextEvent sce){
    //释放连接池
}
```

下面来具体实现。

7.3.2 创建监听器类 AppService

这个类的代码结构如下:

```java
@WebListener
public class AppService implements ServletContextListener {
    public static HikariDataSource HikariCP_POOL;     //外部通过其获取池化连接

    @Override
    public void contextInitialized(ServletContextEvent sce) {
        setupPool();                                   //构建连接池方法
    }
    @Override
    public void contextDestroyed(ServletContextEvent sce) {
        closePool();                                   //释放连接池方法
    }
}
```

7.3.3 连接池的构建和关闭

1. setupPool()

基本上与 7.2.3 节中的代码一样,只是数据源名称改变了一下:

```java
private void setupPool() {
    if (HikariCP_POOL == null) {
        Properties p = new Properties();
        ...
        HikariCP_POOL = new HikariDataSource(config);
    }
}
```

2. closePool()

```java
private void closePool() {
    if (HikariCP_POOL != null) {
        HikariCP_POOL.close();
    }
}
```

7.3.4 修改 DBFactory 类

```java
public class DBFactory {
    private static DBFactory instance;
```

```
            private DBFactory() { super(); }
            public static DBFactory get() {
            if (instance == null) {
                    instance = new DBFactory();
            }
            return instance;
        }
        public Connection getConnection() {
            …
            conn = AppService.HikariCP_POOL.getConnection();
            …
        }
        public void closeConnection(Connection conn) {
            …
        }
}
```

功能更单一，代码更简洁了。

7.4 场景任务挑战——动态配置 HikariCP

目前配置连接池的方法，都是将连接池配置参数写在 Java 程序代码中，例如一开始是写在 DBFactory 中，后来则写在 AppService 的 setupPool() 方法中。这种将配置参数写在 Java 程序代码中的方法，称为硬编码。硬编码的缺陷就是：一旦需要调整参数，例如数据库地址发送改变了、连接池的活动连接数需要调整等，都需要修改程序源代码。

由此看来，需要一种能够动态配置 HikariCP 连接池的方法：当参数需要调整时，不需要修改源代码。有一种处理思路是：使用 properties 资源配置文件，例如在 src/main/resources 下创建 hikariConfig.properties，在这个文件中配置好各种参数。应用启动时，AppService 从 hikariConfig.properties 中读取参数。这个任务，交给读者挑战完成！

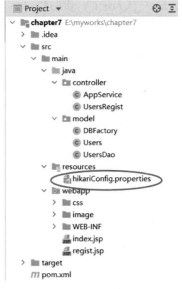

图 7-4 资源配置文件

第 8 章

jQuery前端开发

jQuery 是一个能够简化 JavaScript 编程的 JavaScript 库。本章主要介绍了 jQuery 基础语法、选择器、事件函数、HTML 和 CSS 的处理,以及如何与服务器进行数据交互。熟练掌握 jQuery 的使用,将为我们减少 JS 代码编写量、增强对 HTML 元素的控制能力提供很好的帮助。

8.1 jQuery 概述

8.1.1 jQuery 简介

jQuery 是一个快速、轻量、功能丰富的 JavaScript 库。jQuery 大大简化了对 HTML 文档的各种操作处理,提供了丰富的、跨浏览器的、易于使用的 API。jQuery 的可扩展性、多功能性,改变了 Web 应用中 JavaScript 的编写方式。

jQuery 的口号是"write less,do more",在现实中得到了广泛应用,成为 Web 前端的开发利器。

8.1.2 jQuery 的使用

jQuery 有两种常用的使用方式。

1. 下载 JS 文件

1) 非压缩的 JS 文件

非压缩的 JS 文件名一般叫 jquery-x.x.x.js(x.x.x 是版本号),比压缩版的大不少,一般用于开发环境。官网下载地址是:https://jquery.com/download/,在此使用 3.6.0 这个版本。将其加入 IntelliJ IDEA 开发环境,就能够自动识别 jQuery 的方法函数,为代码编写带来方便性。例如,可将官网下载到的 jquery-3.6.0.js 加入 IntelliJ IDEA 编辑器,如图 8-1 所示。当然,这个步骤并不是必需的。

2) 压缩的 JS 文件

压缩版的 JS 文件名一般用 jquery-x.x.x.min.js,例如 jquery-3.6.0.min.js。这种压缩文件体

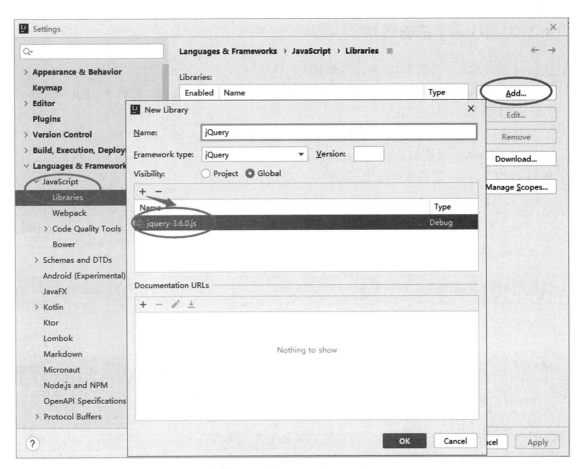

图 8-1　加入 jquery-3.6.0.js

积较小,一般用于生产环境。

将下载到的 JS 文件复制到站点项目的某个文件夹下,例如 js,然后在页面中引用即可:

```
<head><script src="js/jquery-3.6.0.min.js"></script></head>
```

这种方式的好处是由于 jquery-3.6.0.min.js 已经属于站点文件的一部分,所以不需要外部网络的支持。

2. 通过 CDN 引用

jQuery 的 CDN(Content Delivery Network,内容分发网络)主要由 StackPath 网站提供,此外还有谷歌、微软、CDNJS 等。StackPath 引用示例:

```
<script src="https://code.jquery.com/jquery-3.6.0.min.js"
        integrity="sha256-/xUj+3OJU5yExlq6GSYGSHk7tPXikynS7ogEvDej/m4="
        crossorigin="anonymous"></script>
```

这种方式需要外部网络的支持,一旦断网,就会对页面的运行产生影响。

8.1.3 jQuery 基础语法

jQuery 语法形式非常简单,以美元符号打头:
- $("<element></element>")或$("<element/>"):创建新元素,等价于 JS 的 document.createElement("element")。
- $(selector).action():jQuery 基本语法形式。选择符(selector)可用来定位 HTML 元素,action()则表示对定位元素的某种操作。

示例 1:用 jQuery 隐藏页面图片。

```
<img src="image/m0.png" id="logo" width="32" height="32" title="运动">
```

则下面的 jQuery 语句可隐藏该图片:

```
$("#logo").hide();
```

示例 2:用 jQuery 创建示例 1 中的图片。

```
let img = $("<img/>", {
    src: "image/m0.png",
    id: "logo",
    width: "32",
    height: "32",
    title: "运动"
});
```

8.2 jQuery 选择器

选择器的作用,就是获取到指定 HTML 的元素、元素属性或 CSS 属性,以便进行后续的 action()操作。

8.2.1 元素选择器

元素选择器就是选取 HTML 元素,有以下五种典型格式。
- $("img"):选择页面的所有图片。
- $(".message"):选取所有 class="message"的元素。
- $("span.message"):选取所有 class="message"的<span>元素。
- $("#menu"):选取所有 id="menu"的元素。
- $("div#menu"):选取 id="menu"的<div>元素。

在此基础上,jQuery 增强了元素选择器的用法。表 8-1 列出了其他一些常用选择器:

表 8-1 常用选择器

选 择 器	说 明	举例(作用)
:first	第一个元素	$("img:first")(第一张图片)
:last	最后一个元素	$("img:last")(最后一张图片)
:even	所有偶数元素	$("tr:even")(表格的偶数行)
:odd	所有奇数元素	$("tr:odd")(表格的奇数行)
:eq(index)	第 index 个元素	$("select option:eq(0)")(下拉选择框的第一个列表项)
:contains(text)	所有包含 text 的元素	$("span:contains('武汉')")(所有内容中包含"武汉"的<span>元素)
:empty	无子节点的元素	$("td:empty")(表格中内容为空的单元格)
:not(selector)	不为空的元素	$("input.message:not(:empty)")(class="message"中内容不为空的 input 元素)

示例：将页面表格的奇偶行颜色设置为交替变化、单元格为空的背景色设置为红色,如图 8-2 所示。

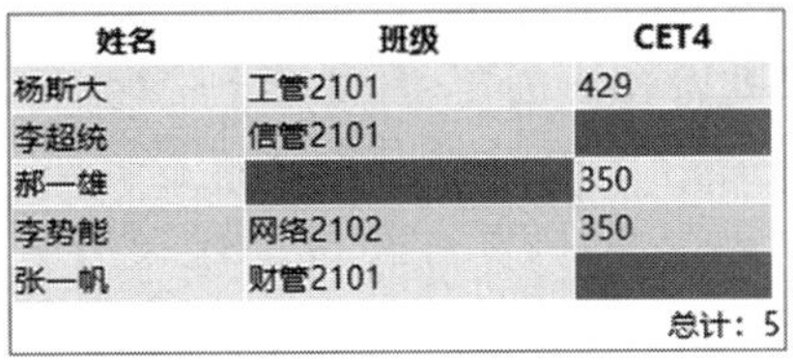

姓名	班级	CET4
杨斯大	工管2101	429
李超统	信管2101	
郝一雄		350
李势能	网络2102	350
张一帆	财管2101	

总计：5

图 8-2 数据表格

为方便理解,这里给出大部分代码:

```
<!DOCTYPE html>
<html lang="zh">
<head>
    <meta charset="UTF-8">
    <title>学生信息查询</title>
    <style>
        table {
            width: 300px;
            margin: auto;
            border: 1px solid #8383e8;
        }
    </style>
    <script src="js/jquery-3.6.0.min.js"></script>
    <script>
        $(document).ready(function () {
            $("tbody tr:odd").css("background", "#e0e0f3");
            $("tbody tr:even").css("background", "#e8f1fb");
            $("tbody td:empty").css("background", "#f00");
        });
    </script>
</head>
```

```
<body>
<table>
    <thead>
        <tr><th>姓名</th><th>班级</th><th>CET4</th></tr>
    </thead>
    <tbody>
        <tr><td>杨斯大</td><td>工管2101</td><td>429</td></tr>
        <tr><td>李超统</td><td>信管2101</td><td></td></tr>
        ……<!-- 类似数据代码,略去 -->
    </tbody>
    <tfoot>
        <tr><td colspan="3" style="text-align: right">总计: 5</td></tr>
    </tfoot>
</table>
</body></html>
```

代码中$(document).ready()使用了jQuery的文档就绪函数,以防止在未完全加载前就运行jQuery代码导致可能出错。这是非常经典的写法,请大家记住!

提示:请先创建好 Maven Webapp 项目 chapter8。与前面章节不同,目前只需要加入 PostgreSQL、Servlet、HikariCP 依赖。另外,从本章开始,页面以 HTML 文件为主,不再采用 JSP 文件。

8.2.2 属性选择器

基于 HTML 元素的属性进行选择。下面列举的是一些常见属性选择器。

◇ $("[属性名]"):选择所有具有指定属性名的元素。例如$("[type]")选择有 type 属性的元素,那么<input type="text" name="username">就会被选择。

◇ $("[属性名=值]"):选择所有属性名等于指定值的元素。例如$("[href='#']")选择全部空链接。

◇ $("[属性名^=值]"):选择所有属性值以指定值开头的元素。例如$("[src^='image']")选择 src 属性值以"image"开头的元素。

◇ $("[属性名$=值]"):选择所有属性值以指定值结尾的元素。例如$("[src$='.png']")选择 src 属性值以".png"结尾的元素。

◇ $(":属性名"):对某些属性值选择的简明写法。例如$(":enabled")选择所有激活的 input 元素,$(":checked")选择那些被选中的复选框,$("div:hidden")则选择全部被隐藏的<div>层。

示例:将友情链接中属于教育机构且名字中含有"科技"二字的,用加粗字体、带点线红色边框突出显示。

```
<script>
    $(document).ready(function () {                      //文档就绪
        $("a[href$='edu.cn']:contains('科技')").css(      //属于教育机构且名字中含有"科技"
            {"border":"1px dotted #f00",                 //点线红色边框
             "font-weight":"bold"});                     //粗体字
```

```
        });
    </script>
    友情链接:<br/>
    <a href = "https://www.hust.edu.cn">华中科技大学</a><br/>
    <a href = "https://www.ifeng.com">凤凰网</a><br/>
    <a href = "https://www.whu.edu.cn">武汉大学</a><br/>
    <a href = "https://www.ebrun.com">亿邦动力网</a><br/>
    <a href = "https://movie.douban.com">豆瓣电影</a><br/>
    <a href = "https://www.uestc.edu.cn">电子科技大学</a>
```

8.2.3 CSS 选择器

在前面的例子中,其实已经使用了 CSS 了。下面主要介绍 jQuery 常用的 CSS 设置与获取方法。

◇ .css(属性名):获取指定 CSS 属性名的属性值,可以是数组表示的多个属性名。

◇ .css(属性名,属性值):设置指定 CSS 属性名的属性值。

◇ .css(属性名,函数):利用函数的返回值来设置 CSS 指定属性名的属性值。这个函数比第 2 个更为灵活。

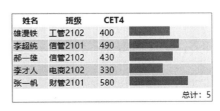

示例:在 8.2.1 节示例的基础上进行改进,在 CET4 后面添加 1 列,并用条形图显示分数情况。当分数大于或等于 425 分时,用绿色显示;否则用红色显示。直方条的长度与 CET4 分数匹配,如图 8-3 所示。

图 8-3 分数条形图

先来看看 HTML 代码:

```
<table id = "cet4">
    <thead>
        <tr><th>姓名</th><th>班级</th><th>CET4</th><th></th></tr>
    </thead>
    <tbody>
        <tr>
            <td>雄漫铁</td><td>工管 2102</td><td>400</td>
            <td><span></span></td>
        </tr>
        ...<!-- 类似数据代码,略去 -->
</table>
```

加粗的代码是增加的列,这里通过设置表格单元格的 width、height、background 来显示直方条。由于 CET4 分数的数值较大,所以用缩小的值"分数 * 0.2"来设置宽度。jQuery 代码如下:

```
$(document).ready(function () {
    $("#cet4 tbody tr:odd").css("background", "#e0e0f3");
    $("#cet4 tbody tr:even").css("background", "#e8f1fb");    //设置 CSS 单个属性值
    $("#cet4 tbody").children().each(function () {            //jQuery 操作,遍历元素
        $(this).children(":last").css("width", function () {  //用函数来设置 CSS 宽度
```

```
                    let w = 0.2 * parseInt($(this).prev().text());   //获取CET4的分数值
                    let color = w<85?"#f00":"#099109";               //判断标准：425×0.2=85
                    $(this).css({                                    //设置CSS多个属性值
                            display: "block",
                            height: "16px",
                            background: color
                    });
                    return w;                                        //返回w作为单元格的宽度
            });
        });
    });
```

在这个例子中，使用了jQuery对元素的操作：遍历<table>表格、获取元素内容（CET4分数）。下面来具体学习jQuery对元素的各种操作。

8.3　jQuery操作

8.3.1　元素操作

表8-2列出了一些常用的元素操作函数（方法）：

表8-2　元素操作

名称	说明	举例（作用）
append()	在被选元素的末尾插入内容	$("div#intro").append("ok");（在id="intro"的<div>层的末尾添加ok） $("div#intro").append($("<b>武汉</b>"));（在intro层末尾添加粗体字"武汉"）
clone()	克隆被选元素	$("div#intro").clone(true);（将intro层克隆一份，true表示深度克隆，即包含层的事件）
html()	获取（设置）被选元素的HTML内容	alert($("div#intro").html());（显示intro层的HTML内容） $("div#intro").html("<img src='image/email.png'>");（将intro层的内容修改成一张图片）
prepend()	在被选元素的开头插入内容	$("div#intro").prepend($("<i>hello,jQuery!</i>"));（在intro层前面插入斜体的文本"hello,jQuery!"）
text()	获取（设置）被选元素的文本内容	$("div#intro").text("ok");（将intro层的内容设置为ok） $("#intro span").text((index, text) => text + "-" + index);（将intro层中每个<span>子元素的内容加上形如"-0""-1"的内容）
insertAfter()/insertBefore()	将内容插入指定元素外部的后面/前面	$("<span/>").text("ok").insertAfter($("#container"));（将内容为ok的<span>元素插入到层外的后面）
empty()	清空被选元素的内容	$("#intro").empty();（将intro元素的内部内容清空）
remove()	删除被选元素	$("#intro").remove();（删除intro元素）
replaceAll() replaceWith()	被选元素全部被替换为指定内容	$("<div>ok</div>").replaceAll("#intro span"); $("#intro span").replaceWith("<div>ok</div>"); （将intro元素内部的<span>全部替换为<div>ok</div>）

示例：动态新增内容。

图 8-4 是一个用于书目录入的界面,显示了一个书名文本框、两个单选按钮。
对应 HTML 代码如下:

```
<div id="books">
    <div>
        <input type="text" size="20">
        <input type="radio" name="r0" value="01" checked/>社会科学
        <input type="radio" name="r0" value="02"/>自然科学
    </div>
</div>
```

现在,需要往界面添加新的一行,用 jQuery 代码处理如下:

```
$(document).ready(() => {
    let newbook = $("#books div:first").clone();
    $("#books").append(newbook);
});
```

这里使用 clone()方法直接复制一份添加到 books 层里面。代码很简单,但是在浏览器中打开页面后效果却有些问题,如图 8-5 所示。

图 8-4　添加一行前　　　　　　　　图 8-5　添加一行后

可以看到,原本每本书后面的单选按钮,现在变成两本书,只能选择一个了！这是因为:单选按钮都需要成组出现。第一组单选按钮 name="r0",则第二组单选按钮的 name 不能还是 r0,例如可改成 name="r1",否则所有单选按钮全部成为一组了,就会出现如图 8-5 所示的不正常情况！要解决这个问题,学习了属性操作就可以实现。

8.3.2　属性操作

jQuery 提供了如下一些对 HTML 元素进行属性操作的方法。

- attr()：获取/设置元素属性值。例如 $("#books div").children(":eq(1)").attr("checked")这句代码,获取到图 8-4 中单选按钮的值为"checked"。
- prop()：跟 attr()类似。例如 $("#books div").children(":eq(1)").prop("checked")返回值是 true。注意,attr()返回字符串型值,而 prop()返回的是逻辑型值。
- removeAttr()：删除所选元素的属性。例如 $("#mylogo").removeAttr("title")删除 mylogo 的 title 属性。
- removeProp()：删除由 prop()设置的属性。例如假定先设置了一个 school 属性 $("#books").prop("school","em"),则可使用 $("#books").removeProp("school")将其删除。

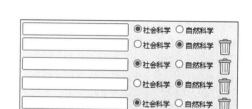

图 8-6 添加后

◇ val()：获取/设置被选元素的值。例如语句 $("#books div").children(":eq(1)").val() 获取到的值为 01。

示例：完善并加强上一节的例子，添加 4 行并交替选中"自然科学"或"社会科学"，同时在后面加一张图片。鼠标移到图片上时，图片提示文字"删除"，鼠标变成手形，如图 8-6 所示。

只需要修改 jQuery 代码如下：

```
for (let i = 1; i < 5; i++) {
    let newbook = $("#books div:first").clone();
    newbook.children(":radio").attr("name", "r" + i);        //重命名 name 属性值
    newbook.children(":eq(1)").prop("checked", i % 2 == 0);  //偶数行选中"社会科学"
    newbook.children(":eq(2)").prop("checked", i % 2 != 0);  //奇数行选中"自然科学"
    newbook.append( $ ("<img/>", {                           //添加图片
        src: "image/delete.png",
        width: "32",
        height: "32",
        title: "删除"   //图片提示文字
    }).css("cursor", "pointer"));                            //鼠标形状设置为手形
    $("#books").append(newbook);
}
```

8.3.3 操作 CSS 类

jQuery 提供了如下给元素动态增删 CSS 类的方法。

◇ addClass()：添加 CSS 类。例如 $("div,table").addClass("menu")。

◇ removeClass()：删除 CSS 类。例如 $("div,table").removeClass("menu")。

◇ toggleClass()：添加或删除 CSS 类。若 CSS 类已经存在，删除；若不存在，则添加。例如 $("div, table").toggleClass("menu")。

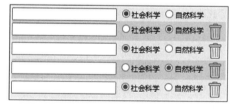

示例：给上一节例子添加奇偶行背景色交替显示，如图 8-7 所示。

先来定义 CSS 样式：

图 8-7 添加 CSS 样式

```
<style>
    #books { /*外层<div>样式*/
        position: absolute;
        top:100px;
        width: 360px;
        margin: auto;
        border: 1px solid #4141f3;
        background: #f3f4f6;
```

```
                font-size: 12px;
                padding: 3px;
            }
            .odd { /* 奇数行样式 */
                background: #dcdcf3;
            }
            .even { /* 偶数行样式 */
                background: #e8f1fb;
            }
            img { /* 删除图片样式 */
                vertical-align: middle;
            }
        </style>
```

接下来,只需要加入一句代码:

```
$("#books div").addClass(index => index % 2 == 0 ? "even" : "odd");
```

这句代码,给 books 层中每个 <div> 子元素添加 CSS 类。通过浏览器可观察到页面执行后文档结构如图 8-8 所示。我们在 addClass() 方法中通过箭头函数处理奇偶行背景色:index 代表每个 <div> 子元素的索引,通过判断 index % 2 == 0 是否是偶数,返回相应的 CSS 类名。

图 8-8 生成的 HTML 元素

8.3.4 遍历操作

遍历,就是对文档中的 HTML 元素或子元素、兄弟元素等进行浏览,下面了解一些常见的遍历方法。

- ◇ children():被选元素的子元素。这个方法前面用过多次了。
- ◇ each():被选元素中的每个元素。在 8.2.3 节的示例中用过。
- ◇ closest():被选元素中的邻近元素。例如将图 8-7 中所有删除图片背景设置为红色,则可这样 $("#books div").children().closest("img").css("background","#f00")。
- ◇ even():偶数位置的元素。例如 $("table").even().css("color","#f00")。
- ◇ odd():奇数位置的元素。例如 $("table").odd().css("color","#f00")。
- ◇ next():下一个兄弟元素。例如 $("#msg").next().text("下一步")。
- ◇ parent():父元素。例如 $(this).parent().tagName 获取当前元素的父元素的标签名。
- ◇ siblings():兄弟元素。例如将 value 值为 02 元素的兄弟元素背景设置为红色则可这样: $("#books div").children("[value='02']").siblings().css("background","#f00")。

示例:将上一节奇偶行设置为交替背景色。

可将这句代码:

```
$("#books div").addClass(index => index % 2 == 0 ? "even" : "odd");
```

用下面的两句代码替换,效果一样:

```
$("#books div").odd().addClass("odd");
$("#books div").even().addClass("even");
```

8.3.5 事件函数

jQuery 提供了丰富的事件函数,用于处理各种事件,例如鼠标单击、移动鼠标等。表 8-3 列出了一些常用事件函数。

表 8-3 常用事件函数

名称	说明	名称	说明
blur()	失去焦点时触发	mousedown()	按下鼠标时触发
change()	发生改变时触发	mouseenter()	鼠标进入时触发
click()	鼠标单击时触发	mouseleave()	鼠标离开时触发
contextmenu()	右击鼠标时触发	mousemove()	鼠标移动时触发
dblclick()	双击鼠标时触发	mouseout()	鼠标移开时触发
event.data	事件返回的数据	mouseover()	鼠标划过时触发
event.target	事件源	mouseup()	松开鼠标时触发
focus()	获得焦点时触发	on()	给被选元素绑定事件
hover()	鼠标悬停时触发	ready()	文档就绪时触发
keydown()	按键按下时触发	select()	文本被选择时触发
keypress()	按键时触发	submit()	提交表单时触发
keyup()	按键松开时触发	serialize()	表单数据序列化为字符串

示例 1:用 jQuery 重写第 4 章 4.7.5 节的文本同步输出功能。

```
$("#txt1").keyup(() => $("#txt2").val($("#txt1").val()));
```

示例 2:单击按钮时,将 8.3.3 节图 8-7 中的全部删除图片替换为另外一张图片 del.png。先放置一个按钮:

```
<button id="btn">单击换图</button>
```

然后用一句 jQuery 代码即可实现换图:

```
$("#btn").on("click", {image: "image/del.png"},
    (event) => $("img").attr("src", event.data.image)
);
```

给按钮绑定 click 事件时,传送图片数据,然后通过 event.data.image 获取该数据,并修改所有图片的 src 属性。

8.4 jQuery 动画

jQuery 提供了基本的动画处理方法。

- fadeIn()/fadeOut()：淡入/淡出元素。例如 $("#books").fadeOut(3000) 表示用 3s 淡出 books 元素。
- fadeToggle()：若已淡出，则 fadeIn()；若已淡入，则 fadeOut()。
- fadeTo()：透明性渐变。例如 $("#books").fadeTo(3000, 0.5) 表示在 3s 内将 books 逐渐变成半透明。
- slideDown()/slideUp()：向下/向上滑动元素。例如 3s 内缓慢下拉显示 books 层 $("#books").css("display","none").slideDown(3000)。
- animate({params}[, speed])：自定义动画。params 设置参数，speed 控制速度。

示例 1：对 8.3.3 节的 books 层进行动画处理。单击按钮，3s 内收缩至消失。

先在页面放置一个按钮：

```html
<button id="btn">单击试试</button>
```

然后编写 jQuery 代码：

```javascript
$("#btn").on("click", () => {
        $("#books").animate({
            width: "hide",
            height: "hide"
        }, 3000);
    }
);
```

示例 2：在 3s 内向右移动 books 层 100 像素到新位置，并将透明度设置为 0.2，上下边框加粗至 3px。

```javascript
$("#books").animate({
    left: "+=100px",
    opacity: 0.2,
    borderTopWidth: "3px",
    borderBottomWidth: "3px"
}, 3000);
```

8.5 与服务器交互

jQuery 提供了几种简便方法，方便与后台服务器进行数据的交互处理。通常，后台数据以 JSON 格式发送到前端。Jackson 提供了强大的后台 JSON 数据处理能力。

8.5.1 用 Jackson 格式化数据

1. Jackson 简介

Jackson 是流行的基于 Java 的 JSON 处理框架。占用内存低、同类框架中性能强劲、社区非常活跃、Spring 官方推荐等特点，使得 Jackson 应用非常广泛。

Jackson 可实现 Java 对象与 JSON 对象互相转化，即将 Java 对象转成 JSON 对象，或将 JSON 对象转成 Java 对象。

2. 加入 Jackson 依赖

在项目的 pom.xml 中加入以下代码并更新：

```xml
<dependency>
    <groupId>com.fasterxml.jackson.core</groupId>
    <artifactId>jackson-databind</artifactId>
    <version>2.12.4</version>
</dependency>
```

3. Jackson 常用方法

◇ new ObjectMapper()：创建 Jackson 对象。

```
ObjectMapper om = new ObjectMapper();
```

◇ om.writeValueAsString(Object value)：将 Java 对象转换为 JSON 字符串。

```
String jStr = om.writeValueAsString(list);
```

◇ readValue(String content，Class<T> valueType)：将 JSON 字符串转换为 Java 对象。

```
String usr = "{\"username\":\"张丰\",\"password\":\"252689\"}";
Users user = om.readValue(usr, Users.class);
```

◇ readTree(BufferedReader reader)：解析 JSON 字符串，返回 JsonNode 对象。

```
JsonNode jn = om.readTree(request.getReader());
String username = jn.get("username").asText();
String password = jn.get("password").asText();
```

8.5.2 $.ajax

进行异步的 HTTP 请求。关于异步处理，请参阅第 4 章 4.8 节的内容。$.ajax 常用语法格式如下：

```
$.ajax({
    type: method,                                          //数据提交模式,常用的有 get 或 post
    url: url,                                              //请求地址
    data: data,                                            //要提交的数据
    contentType: dataType,                                 //内容编码类型
    success: function (data, textStatus, jqXHR) {          //提交成功的回调函数
        //这里,可对获取到的后台数据 data 进行处理
    },
    error: function (data, textStatus, errorThrown) {      //提交失败的回调函数
        //这里,可进行出错处理
    }
})
```

或者,采用链式调用的写法:

```
$.ajax({
    type: method,
    url: url,
    data: data,
    contentType: dataType
}).done((data, textStatus, jqXHR) => {          //可对获取到的后台数据 data 进行处理})
.fail((data, textStatus, errorThrown) =>{       //可进行出错处理})
.always(function (data, textStatus, jqXHR) {    //可进行必须执行的操作});
```

textStatus 表示后台返回的 HTTP 响应状态码,例如状态码 200 表示成功、404 表示页面找不到。jqXHR 则用于获取服务器响应返回的 HTTP 头信息,例如图 8-9 显示了某次登录请求的服务器响应信息。若要获取 Authorization、Content-Type 的值,则可这样:

```
let auth = jqXHR.getResponseHeader("Authorization");
let contentType = jqXHR.getResponseHeader("Content-Type");
```

图 8-9　HTTP 响应头

示例：调用后台 users.list 并传送数据 001 进行查询，将后台返回的全部记录在控制台日志输出。

```
$.ajax({
    type: "get",                                    //get 模式提交
    url: "users.list",
    data: '{"t":"001"}',                            //JSON 字符串
    contentType: "application/json",                //JSON 数据
    success: function (data) {
            data.forEach(user => {                  //data 中的 JSON 对象
                for (let key in user) {             //遍历对象的每个属性
                    console.log(user[key]);         //输出对象的每个属性值
                }
            });
    },
    error: function (data) {
        alert("出错!")
    }
})
```

8.5.3 $.get 和 $.getJSON

基于 HTTP GET 请求模式，通过指定 URL 向服务器发送数据，并可获取后台数据。基本使用格式如下：

```
$.get(url)  //可换成 $.getJSON
    .done(function (data, textStatus, jqXHR) {
        //可对获取到的后台数据 data 进行处理
    })
    .fail(function (data, textStatus, errorThrown) {
        //可进行出错处理
    })
    .always(function (data, textStatus, jqXHR) {
        //可进行必须执行的操作
    });
```

$.getJSON 要求后台返回的 data 是 JSON 格式，不然出错，而 $.get 则没有这个要求。

示例：用 $.getJSON 重写上一节的示例。

```
$.getJSON("users.list?t = '001'") //调用后台 users.list 并传送数据 001
    .done(data => {
        data.forEach(user => {
            for (let key in user) {
                console.log(user[key]);
            }
        });
    })
    .fail(() => alert("出错!"));
```

8.5.4 $.post

跟$.get有些类似,不过是以HTTP POST方式向服务器发送数据。语法格式如下:

```
$.post(url,data)
    .done(function (data, textStatus, jqXHR) {……})
    .fail(function (data, textStatus, errorThrown) {……})
    .always(function (data, textStatus, jqXHR) {……});
```

示例:用jQuery改写第6章6.3节的用户注册示例。
(1) 新建regist.html,将原regist.jsp主要内容复制到regist.html中,主要HTML代码如下:

```html
<div id="registForm" class="user-regist">
    注册教务后台管理通行证<br>
    <img src="image/username.png">
    <input type="text" class="rinput" placeholder="用户名" id="usr">
    <br><img src="image/password.png">
    <input type="password" class="rinput" placeholder="密　码" id="pwd">
    <br><img src="image/repwd.png">
    <input type="password" class="rinput" placeholder="确认密码" name="rpwd">
    <br><img src="image/email.png">
    <input type="email" class="rinput" placeholder="email" id="email">
    <br><input type="radio" name="role" value="student" checked>学生
    <input type="radio" name="role" value="teacher">教师<br>
    <button type="button" class="rbutton" id="submitBtn">提交注册</button>
    <br>
    <span id="message"></span>
</div>
```

这里有几个变化:首先,没有使用表单,而是使用了<div>层;其次,<input>输入域使用了id属性(包括提交注册按钮),方便jQuery选取这些元素;最后,使用<span>显示注册成功与否的消息。
(2) regist.html的JS代码。

```javascript
<script src="js/jquery-3.6.0.min.js"></script>
<script>
    $(document).ready(() => {
        $("#submitBtn").click(() => {
            $.post("regist.do", {                    //提交给regist.do的JSON对象数据
                username: $("#usr").val(),
                password: $("#pwd").val(),
                email: $("#email").val(),
                role: $('input[name="role"]:checked').val()})
            .done(data => $("#message").text(data))  //显示regist.do返回的数据
            .fail(() => $("#message").text("出错,无法提交注册!"));
        }
    );
```

```
        });
    </script>
```

（3）修改 Servlet 类 UsersController 的 doGet()。

```
@Override
protected void doGet(HttpServletRequest request, HttpServletResponse response) throws IOException {
    response.setContentType("text/html; charset=UTF-8");
    PrintWriter out = response.getWriter();
    ObjectMapper om = new ObjectMapper();                                   ①
    String str = om.writeValueAsString(request.getParameterMap());          ②
    String jstr = str.replaceAll("\\[", "").replaceAll("\\]", "");          ③
    Users user = om.readValue(jstr, Users.class);                           ④
    if (new UsersDao().isExisted(user.getUsername()))
         out.print("该用户名已被注册!");
    else if (new UsersDao().addUser(user))
         out.print("注册成功!");
    else
         out.print("出错,注册失败!");
}
```

说明：

① 创建一个 Jackson 对象 om，以便下面进行数据转换处理。

② 前端 jQuery 传送 JSON 对象数据后，这里用 request.getParameterMap()获取键值对数据 Map，再使用 Jackson 的 writeValueAsString()方法，转换成 JSON 字符串。转换后的字符串形式类似于：

```
{"username":["张无双"],"password":["666"],"email":["zws@126.com"],"role":["teacher"]}
```

可以看出，转换后的属性值用的是数组，并不完全吻合我们的使用要求。

③ 由于②中转换的属性值为数组，因此需要去掉多余的字符"["""]"。这里使用方法 replaceAll()将其都替换为空字符，现在 jstr 的内容吻合我们的要求了：

```
{"username":"张无双","password":"666","email":"zws@126.com","role":"teacher"}
```

④ 将 JSON 字符串 jstr 转换成 Java 对象 user。

注意：使用 jQuery 后，不再适合用第 6 章中的 RequestDispatcher 进行页面转发了，所以这里只利用 out 向前端返回注册成功与否的字符串数据。

（4）修改 UsersDao 类的 addUser()方法。

主要是将该方法的返回值修改为 boolean 型：

```
public boolean addUser(Users user) {
    boolean sucess = false;
    Connection conn = null;
    ...
                ps.close();
```

```
                sucess = true;
            } catch (Exception e) {
                e.printStackTrace();
            }
            ...
            return sucess;
        }
```

8.6 场景应用示例

8.6.1 下拉选择框联动

1. 应用需求

用下拉选择框实现学院、专业名称之间的联动,如图 8-10 所示。当选择不同院系时,专业名称发生相应变化。

图 8-10 院系专业联动

2. 处理思路

每个院系用一个 JSON 对象表示,多个院系则用数组存放。当选择不同的院系时,通过 change 事件的 event 事件源 event.target 获取到当前院系列表框的索引,然后通过该索引获取对应数组对象的专业名称。

3. 代码实现

```
<script src="js/jquery-3.6.0.min.js"></script>
<script>
    $(document).ready(function () {
        const schools = [
            {name:"管理学院", majors:["信息管理","电子商务","物流管理","工商管理"]},
            {name:"经济学院", majors:["国际贸易","金融学","财政学"]},
            {name:"传媒学院", majors:["动画","广告","数字媒体","新闻传播"]}
        ];
        //构建院系下拉选择框
        schools.forEach(s => $("#departs").append($("<option/>").text(s.name)));
        //构建专业下拉选择框
        $("#departs").change(event => {                          //院系下拉变化时
            $("#majors").empty();                                //清空旧数据
            schools[event.target.selectedIndex].majors.forEach(  //遍历当前院系的专业名称
                m => $("#majors").append($("<option/>").text(m)) //构建专业名称
            );
        });
        $("#departs").change();                                  //构建默认的院系专业列表框
    });
</script>
院系<select id="departs"></select>
专业<select id="majors"></select>
```

读者可以更进一步实现：院系专业数据，来自数据库！从上述代码来看，用包含 JSON 对象的数组处理这个问题，简洁很多。网上及有些辅导书中，不少人用多个数组来实现这个功能，复杂了不少。

视频讲解

8.6.2 学生信息查询

如果已经完成了第 6 章 6.4 节的场景任务挑战，应该已经创建好了 student 数据表，可用来进行信息查询。

1. 应用需求

输入学生姓名，查询出全部同名学生信息，如图 8-11 所示。若查询不到数据，则给予相应提示信息。

图 8-11 学生信息查询

2. 处理思路

查询架构设计如图 8-12 所示。这里为什么增加了一个 StudentService 类？是为了将 StudentController 控制器与业务处理逻辑类 StudentDao 解耦，也就是说控制器只与服务类 StudentService 打交道，类似于顾客只与销售部门打交道，并不与生产部门打交道。类的数量是增加了，但管理上更清晰了。

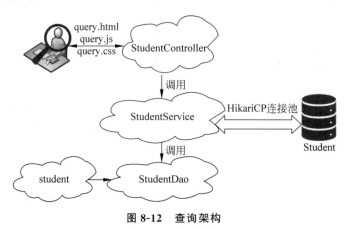

图 8-12 查询架构

从技术的角度看,这个查询主要涉及数据集合的处理。后台用 List 存放查询到的多条记录,并转换成 JSON 字符串发送给前端的 jQuery。前端创建<table>表格并用 forEach 遍历 List 显示出全部数据。

3. 前端:HTML 页面 query.html

```html
<link rel="stylesheet" type="text/css" href="css/querystu.css"/>
<script src="js/jquery-3.6.0.min.js"></script>
<script src="js/query.js"></script>
<script>
    $(document).ready(() => {
        $.doQuery();          //调用 query.js 中的 doQuery()函数
    });
</script>
<div>
    <img src="image/username.png" height="32" width="32"/>
    <input name="sname" id="sname" type="text" placeholder="请输入姓名..."/>
    <input id="queryBtn" type="button" value="查 询"/>
</div>
<div id="msgdiv">
    <span id="message"></span>
</div>
<div id="studiv"></div>
```

与前面不同,这里将 jQuery 代码放置在单独的 JS 文件 query.js 中以方便管理,页面代码简洁不少,只需要调用 query.js 中定义的查询函数 doQuery()即可。

页面放置了两个层:msgdiv 层用于显示未查询到或出错的信息提示;studiv 用于显示查询到的学生记录。这两个层,一开始都通过 CSS 设置"display:none"为不显示。

页面前面链接的样式 css/querystu.css,代码略去,留给读者补充完成。

4. 前端:query.js

```javascript
$.doQuery = () => {
    $("#queryBtn").on("click", () => {
        $.getJSON("query.do", {sname: $("#sname").val()})     //传送待查询姓名
            .done(data => {
                if (data.length == 0) {
                    $("#message").text("抱歉,未查询到相关记录!");
                    $("#studiv").fadeOut(2000);          //淡出 studiv 层
                    $("#msgdiv").fadeIn(3000);           //淡入 msgdiv 层
                    return;
                }
                $("#studiv").empty();                    //清空旧数据
                let th = ["学号","姓名","班级","电话","地址","邮编"];
                let table = $("<table/>");
                let tr = $("<tr/>");
```

```javascript
                th.forEach(h => {                      //构建表头
                        tr.append($("<th/>").text(h));
                });
                table.append(tr);
                data.forEach(student => {
                        let tr = $("<tr/>");
                        for (let key in student) {     //遍历每个学生对象的属性
                                tr.append($("<td/>").text(student[key]));    ①
                        }
                        table.append(tr);
                });
                $("#studiv").append(table);
                $("#msgdiv").fadeOut(2000);
                $("#studiv").fadeIn(3000);
            })
            .fail(() => $("#message").text("出错,无法查询!"));
        }
    );
}
```

说明:

① 将属性值填充到表格的单元格里面。循环中的 key 是指属性名(例如 sno), student[key]则是指属性值(例如 004004)。

5. 后端: Student 实体类

```java
public class Student implements Serializable {
        private String sno;
        private String sname;
        private String sclass;
        private String tel;
        private String address;
        private String postcode;
        public String getSno() {
            return sno;
        }
        public void setSno(String sno) {
            this.sno = sno;
        }
        ...    //其他 setter、gettter 方法,略去
}
```

6. 后端: StudentDao 业务逻辑处理类

```java
public class StudentDao {
        public List<Student> getStudentsByName(String sname) {
```

```java
        Connection conn = null;
        List<Student> list = new ArrayList<>();
        try {
            conn = DBFactory.get().getConnection();
            String sql = "select * from student where sname = ?";
            PreparedStatement ps = conn.prepareStatement(sql);
            ps.setString(1, sname);
            ResultSet rs = ps.executeQuery();
            while (rs.next()) {
                Student s = new Student();
                s.setSno(rs.getString("sno"));
                s.setSname(rs.getString("sname"));
                s.setSclass(rs.getString("sclass"));
                s.setTel(rs.getString("tel"));
                s.setAddress(rs.getString("address"));
                s.setPostcode(rs.getString("postcode"));
                list.add(s);
            }
            rs.close();
            ps.close();
        } catch (Exception e) { e.printStackTrace(); }
        finally { DBFactory.get().closeConnection(conn); }
        return list;
    }
}
```

getStudentsByName()方法返回 List 列表，列表中存放的是 Student 对象。

7. 后端：StudentService 服务类

```java
public class StudentService {
    private static StudentService instance;
    private StudentService() { super(); }
    public static StudentService get() {
        if (instance == null) {
            instance = new StudentService();
        }
        return instance;
    }
    public List<Student> queryStudentsByName(String sname) {
        return new StudentDao().getStudentsByName(sname);
    }
}
```

该类负责与业务逻辑处理类 StudentDao 打交道，其中 instance 是一个静态的类实例变量。StudentService 类对外暴露 queryStudentsByName()方法以供使用。

8. 后端：StudentController 控制器类

```java
@WebServlet(name = "StudentController", value = "/query.do")
public class StudentController extends HttpServlet {

    @Override
    protected void doGet(HttpServletRequest request, HttpServletResponse response) throws IOException {
        request.setCharacterEncoding("utf-8");
        response.setContentType("text/html; charset=UTF-8");
        String sname = request.getParameter("sname");
        List<Student> students = StudentService.get().queryStudentsByName(sname);
        ObjectMapper om = new ObjectMapper();
        response.getWriter().print(om.writeValueAsString(students));
    }

    @Override
    protected void doPost(HttpServletRequest request, HttpServletResponse response) throws IOException {
        doGet(request, response);
    }
}
```

8.7 场景任务挑战——动态增删书目

在第 4 章 4.12 节成功用 JavaScript 实现了动态增删书目的场景应用。现在，请读者用 jQuery 来完成第 4 章图 4-14 所展示的动态增删书目功能！

Spring Boot开发基础

Spring Boot 在 Java Web 开发领域是一个重量级的存在。本章主要介绍了 Restful、响应式处理、Spring WebFlux、R2DBC、Hazelcast 分布式内存网格等内容。熟练掌握 Spring Boot 相关技术，将为我们高质量、高效率地开发 Web 应用系统提供坚实的技术台阶。

9.1 RESTful 概述

9.1.1 REST 简介

REST 是 REpresentational State Transfer（表述性状态传递）的缩写，由 Adobe 首席科学家、Apache 联合创始人、HTTP 协议作者之一的 Roy Fielding 于 2000 年在其论文中首次提出。Roy Fielding 将 REST 定义为一种混合的、有约束的、基于网络的体系结构样式，后来逐渐成为 Web 应用程序的设计风格和开发方式。

REST 主要对 Web 访问资源的 URL 进行了规范，具体来说就是对 Web 请求的以下四种方式进行规范。

◇ GET：访问或查看资源。
◇ POST：提交数据，新建一个资源。
◇ PUT：发送数据，更新资源。
◇ DELETE：删除资源。

这四种方式分别对应了数据库 CRUD 操作中的 select、insert、update、delete 处理。

9.1.2 RESTful 要义

遵守 REST 原则的 Web 服务风格，被称为 RESTful。

RESTful 对 Web 资源访问进行了统一限定，以便通过统一的接口形式适应不同平台。一般使用 HTTP POST/GET 进行数据交互，而请求、应答数据格式一般为 JSON 格式。

9.1.3 RESTful 请求风格

1. 与传统请求风格的比较

先来看看传统风格的 Web 请求路径。

```
http://localhost:8080/em/query.do?tag=01&pro=0704
```

下面详细解读这条 Web 请求路径。
- 请求协议：http   请求模式：GET
- 请求域名：localhost   请求端口：8080
- 请求路径：em/query.do   请求参数串：tag=01&pro=0704

RESTful 风格的请求路径则是这样的：http://localhost:8080/em/01/0704

2. RESTful 请求路径格式

路径格式为：{变量名[:正则表达式]}，其中正则表达式为可选。下面列出了一些请求路径的可能形式及示例。

- /student/{id}

  http://localhost:8080/em/student/106、http://localhost:8080/em/student/021

- /{name}-{zip}

  http://localhost:8080/em/wuhan-027、http://localhost:8080/em/Hangzhou-0571

- /admin/{action}/user

  http://localhost:8080/em/admin/add/user、http://localhost:8080/em/admin/del/user

Spring Boot 对 RESTful 风格提供了良好支持！从本章开始，在页面各种处理中使用 Restful 路径请求风格。

9.2 Spring Boot 概述

9.2.1 Spring Boot 简介

Spring Boot 是由 Pivotal 团队设计提供的 Java 平台上的一种开源应用框架，致力于快速、高质量地构建应用系统。

Spring Boot 以其灵活、安全、广泛支持的特性，被誉为 Java Web 开发领域的"王者"。Spring Boot 可帮助我们轻松构建独立的、生产级的、基于 Spring 的应用程序。Spring Boot 通过在 Maven 项目的 pom.xml 文件中添加依赖，使得开发人员开箱即用，无须关注烦琐的依赖项管理工作，而是专注业务逻辑的处理即可。

Spring Boot 能够提高开发的快速性、质量的一致性，紧密跟随技术趋势的发展，非常值得深入学习。

9.2.2 创建 Spring Web MVC 项目

1. 创建 chapter9 项目

利用 Spring Initializr 创建基于 Spring Web MVC 框架的项目。

(1)利用 IntelliJ IDEA 的 New→Project 菜单,弹出 New Project 对话框。选择 Spring Initializr,输入项目名 chapter9,并选中 Maven、War(Web 压缩文件)等,如图 9-1 所示。选择 War 的目的是以后可以很方便地发布到外部 Tomcat 服务器。

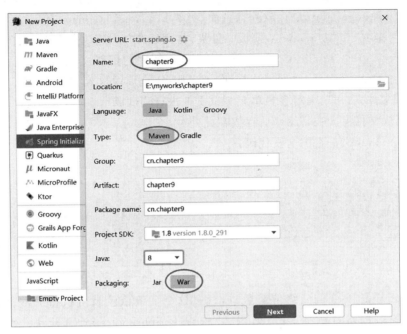

图 9-1　新建 chapter9 项目

(2)单击 Next 按钮,再选中 Spring Web 复选框,单击 Finish 按钮,完成创建,如图 9-2 所示。

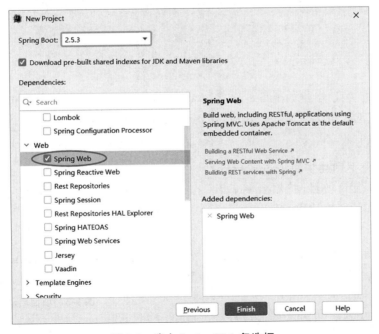

图 9-2　选中 Spring Web 复选框

2. 项目结构

创建好的 chapter9 项目结构主要包括以下部分。

◇ src/main/java：存放 Java 程序文件。

◇ src/main/resources/static：存放静态页面，例如 html、css、js、image 等。

◇ src/main/resources/templates：模板文件夹，常用于存放动态文件（我们暂不需要）。

◇ src/main/resources/application.properties：应用程序的配置文件，可在该文件中配置诸如项目服务器地址、端口号、数据库连接池、消息服务、全局常量等。

◇ test：测试文件夹，用于编写各种单元测试文件进行测试。

◇ target：目标文件夹。项目编译后存放在此处。

再来看看 pom.xml 文件中的这段代码：

```xml
<parent>
    <groupId>org.springframework.boot</groupId>
    <artifactId>spring-boot-starter-parent</artifactId>
    <version>2.5.3</version>
    <relativePath/>
</parent>
```

代码定义了 Spring Boot 的父级依赖版本，通过这个父级依赖就可执行一些插件的自动化配置、自动化的资源过滤等。例如，对 artifactId 中以 spring-boot-starter 开头（例如 spring-boot-starter-web）的依赖项自动配置合适的版本，避免出现版本冲突问题。

而下面的配置代码：

```xml
<groupId>cn.chapter9</groupId>
<artifactId>chapter9</artifactId>
<version>0.0.1-SNAPSHOT</version>
<packaging>war</packaging>
```

则定义了项目相关信息，例如项目名称、版本、打包方式等。一般有两种打包方式：war 和 jar。按照这个配置，生成的打包文件为：chapter9-0.0.1-SNAPSHOT.war。

3. 启动项目

在 static 文件夹下创建主页 index.html，输入"我的主页"。启动项目，控制台将输出如图 9-3 所示内容，并输出类似于 Started Chapter9Application in 4.957 seconds（JVM running for 6.819）字样的信息，表示成功启动。

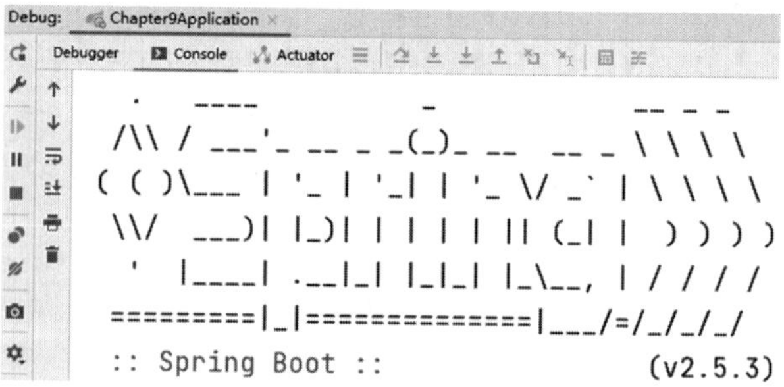

打开浏览器，地址栏输入 http://localhost:8080，将打开主页 index.html 显示相应内容。

图 9-3 项目启动

9.2.3 Spring Boot 入口类

自动生成的入口类有 2 个：Chapter9Application 和 ServletInitializer。
- Chapter9Application 类：利用注解@SpringBootApplication 定义了该类是 Spring Boot 应用程序，并启用自动配置处理。
- ServletInitializer 类：配置使用外部 Servlet 容器（例如 Tomcat），设置应用源为 Chapter9-Application 类。

这两个类可以合并成一个类：

```java
@SpringBootApplication
public class Chapter9Application extends SpringBootServletInitializer {
    public static void main(String[] args) {
        SpringApplication.run(Chapter9Application.class, args);
    }
    @Override
    protected SpringApplicationBuilder configure(SpringApplicationBuilder application) {
        return application.sources(Chapter9Application.class);
    }
}
```

9.2.4 配置 HikariCP 连接池

1. application 配置文件

项目 src/main/resources 文件夹可放置各类配置文件，例如应用配置 application.properties 或其他配置文件。现在，在 application.properties 文件中配置将在 chapter9 项目中使用的 HikariCP 连接池：

```
spring.datasource.url = jdbc:postgresql://localhost:5432/tamsdb
spring.datasource.username = admin
spring.datasource.password = 007
spring.datasource.hikari.maximum-pool-size = 20
spring.datasource.hikari.minimum-idle = 5
spring.datasource.hikari.pool-name = HikariCP
spring.jpa.open-in-view = false
```

最后一句代码禁止在视图渲染时执行数据库查询。也可以不使用 application.properties，而是使用 application.yml 配置文件：

```yaml
spring:
  datasource:
    url: jdbc:postgresql://localhost:5432/tamsdb
    username: admin
    password: '007'
```

```
hikari:
    maximum-pool-size: 20
    minimum-idle: 5
    pool-name: HikariCP
jpa:
    open-in-view: false
```

相较于前者,application.yml 可读性更强。

2. 加入 PostgreSQL 驱动依赖

在 pom.xml 中加入:

```xml
<dependency>
    <groupId>org.postgresql</groupId>
    <artifactId>postgresql</artifactId>
    <version>42.2.23</version>
</dependency>
```

9.2.5 Spring Boot 常用注解

Spring Boot 提供了丰富的注解,对类、方法进行功能性标注。表 9-1 列出了一些常用注解。

表 9-1 常用注解

注解	说明	注解	说明
@Service	业务层处理	@Autowired	自动注入
@Repository	数据访问处理	@Id	主键
@Component	常规组件	@Entity	实体类
@RestController	控制器	@Table	实体类映射的数据表
@RequestMapping	请求的映射地址	@Column	实体类属性与表对应的字段
@PostMapping	Post 映射地址	@Value	配置属性的属性值
@GetMapping	Get 映射地址	@RequestBody	前端请求体中的 JSON 数据
@Configuration	配置	@PathVariable	请求路径变量
@EnableWebFlux	启用 WebFlux	@Query	标注查询语句
@EnableScheduling	启用计划任务	@Param	查询语句中的参数
@Scheduled	计划任务	@Modifying	更新操作
@Bean	业务对象	@Transactional	启用事务处理

示例 1:标注服务类 UsersService。

```
@Service
public class UsersService {
    ...
}
```

示例 2：标注数据访问类 UsersDao。

```
@Repository
public class UsersDao {
    public Users getUser(Users user) {
        ...
    }
}
```

示例 3：标注控制器类 StudentController。

```
@RestController
@RequestMapping("stu")
public class StudentController {
    @GetMapping("query/{sname}")
    public String queryByName(@PathVariable String sname) {
        ...
    }
    @PostMapping("save")
    public String saveStudent(@RequestBody Student student) {
        ...
    }
}
```

@RestController 标注 StudentController 类为控制器，该类的请求 URL 映射为 stu。@GetMapping("query/{sname}")标注 queryByName()方法需要用 Get 模式访问，其映射地址为 query/{sname}，这个地址里面包含了路径参数 sname。假定传递给 sname 的数据为 tom，则访问 queryByName()方法的地址类似于：

```
http://localhost:8080/stu/query/tom
```

而前端页面访问 saveStudent()方法的方式是（以 jQuery 为例）：

```
$.post("stu/save",mydata)
```

前端页面可传递一个 Student 类型的 JSON 字符串 mydata，@RequestBody 注解则标注接收 mydata 数据，随后转换成 student 对象，无须编写额外的转换代码，非常方便。下一节将用示例演示其应用。

9.2.6　JpaRepository 数据访问

1. JPA 简介

JPA 是 Java Persistence API 的简称，负责将运行期的实体对象持久化到数据库中。所谓持久化，通俗地说就是保存到数据库中。Spring Boot 通过 Spring Data JPA 提供了对数据库数据的处理支持。可利用封装后的 JpaRepository，大大简化对数据库的访问处理。

2. 加入 JPA 依赖

在 pom.xml 中加入：

```xml
<dependency>
    <groupId>org.springframework.boot</groupId>
    <artifactId>spring-boot-starter-data-jpa</artifactId>
</dependency>
```

3. 数据访问接口 JpaRepository

Spring Boot 提供了数据访问接口 JpaRepository，该接口提供了丰富的用于数据库的增、删、改、查方法。只需要定义自己的接口类，扩展自 JpaRepository 即可：

```java
public interface UsersRepository extends JpaRepository<Users, String> {
    …
}
```

代码用泛型限定了接口操作的实体类型为 Users 及实体类中的关键字类型为 String。

4. 使用 JPQL

JPQL（Java Persistence Query Language，持久性查询语言）是 Spring 提供的一种面向对象的查询语言，即通过实体对象来操作 SQL 语句（例如 Hibernate，这不是本书讨论内容）。JPQL 简单、强大，可单值或多值查询数据，也可更新数据。JPQL 语法结构与 SQL 语句类似，只不过前者操作的是实体对象，而后者操作的是数据表。

JpaRepository 为了方便处理，使用了一系列关键词来规范并简化 JPQL 的数据访问操作。表 9-2（以处理 Student 数据为例）列出了部分常用的命名规范用例。

表 9-2 命名规范用例

示　　例	JPQL
Student **getBy**Sno(String sno)	select s from Student s where s.sno=?1
List<Student> **findDistinctBy**Sname(String sname)	select distinct s from Student s where s.sname=?1
List<Student> **findBy**Sname(String sname)	…where s.sname=?1
Student **findBy**Sname**And**Tel(String sname,String tel)	…where s.sname=?1 and s.tel=?2
Student **findBy**Sname**Or**Tel(String sname,String tel)	…where s.sname=?1 or s.tel=?2
List<Student> **findBy**Sno**After**	…where s.sno>?1
List<Student> **findBy**Sclass**In**(List<String> sclass);	…where s.sclass in ?1
Student **findBy**Sname**OrderBy**Sno**Desc**(String sname)	…where s.sname =?1 order by s.sno desc
List<Student> **findBy**Sname**Like**(String sname)	…where s.sname like ?1
List<Student> **findAllBy**Sname(String sname)	…where s.sname=?1
Student **findStudentBy**Sname**And**Tel(String sname, String tel)	…where s.sname=?1 and s.tel=?2
void **deleteBy**Sno(String sno)	delete from Student where sno=?1
void **delete**Student**By**Sname**And**Tel(String sname, String tel)	…where sname=?1 and tel=?2
void **saveAndFlush**(student)	insert 操作（数据库中不存在该记录）或 update（数据库中已存在同样的记录）

示例1：查询 users 表中指定用户名、密码的记录。

```
public interface UsersRepository extends JpaRepository<Users, String> {
    Users findByUsernameAndPassword(String username,String password);
}
```

示例2：查询 student 表中指定若干班级的学生记录。

```
public interface StudentRepository extends JpaRepository<Student, String> {
    List<Student> findBySclassIn(List<String> sclass);
}
```

5. 自定义查询@Query

有时关键词查询方式并不能完全满足查询需求,或者希望能够掌控查询语句,这时就需要通过命名查询@Query 来自定义 JPQL 语句。下面通过一些示例来学习其使用方法。

示例1：根据学号查询学生信息。

```
@Query("select s from Student s where s.sno = ?1")
Student findBySno(String sno);
```

特别值得注意的是,@Query 中并不是 SQL 语句,而是 JPQL 语句,所以 Student 不能写成 student,因为这里查询的是实体对象而非数据库的数据表。

示例2：查询指定姓名、指定班级的学生信息。

```
@Query("select s from Student s where s.sname = ?1 and sclass = ?2")
List<Student> findBySnameAndSclass(String sname,String sclass);
```

示例3：查询指定姓名、指定班级的学生信息。

```
@Query("select s from Student s where s.sname = :sname and sclass = :sclass")
List<Student> findBySnameAndSclass(@Param("sname") String sname,
                                    @Param("sclass") String sclass);
```

跟示例2有差别,这里使用了命名参数。

示例4：查询姓名中包含指定字符的学生记录。

```
@Query("select s from Student s where s.sname like %:sname%")
List<Student> findBySnameWithSpelExpression(@Param("sname") String sname);
```

当然,也可以使用示例5中原生的 SQL 语句。

示例5：用 SQL 语句、根据学号查询学生信息。这里可以用小写的 student。

```
@Query(value = "select * from student where sno = ?1 order by sno desc", nativeQuery = true)
Student findBySno(String sno);
```

如果要对记录进行更新或删除操作,则需要加上@Modifying注解。

示例6:修改Users表中指定用户的密码。

```
@Modifying
@Query("update Users u set u.password = ?1 where u.username = ?2")
int updateUsersPassword(String password, String username);
```

示例7:删除student表中指定学号的记录。

```
@Modifying
@Query("delete from Student s where s.sno = ?1")
void deleteStudentBySno(String sno);
```

示例8:向Users表中插入一条指定用户名、密码的新记录。

```
@Modifying
@Query(value = "insert into users values(?1, ?2)",nativeQuery = true)
void insertUser(String username,String password);
```

6. 使用JPA实现用户注册

我们在第8章的8.5.4节实现了用户注册功能。现在用JpaRepository来重新实现,比较两者看有哪些变化,体会JpaRepository的用法。

1)项目结构

图9-4是完成用户注册功能后的项目结构。显然,项目的结构发生了变化,请事先创建好相应的结构。

2)前端:修改jQuery代码

HTML代码无须修改,修改后的jQuery代码如下:

图9-4 项目结构

```
<script src = "js/jquery-3.6.0.min.js"></script>
<script src = "js/core.min.js"></script>
<script src = "js/lib-typedarrays.min.js"></script>
<script src = "js/md5.min.js"></script>                                    ①
<script>
    $(document).ready(() => {
        $("#submitBtn").click(() => {
            $.ajax({
                type: 'post',
                url: 'usr/regist',                                          ②
                data: JSON.stringify(                                       ③
                    {
                        username: $("#usr").val(),
```

```
                            password: String(CryptoJS.MD5($("#pwd").val())),    ④
                            email: $("#email").val(),
                            role: $('input[name = "role"]:checked').val()
                        }
                    ),
                    contentType: "application/json"                              ⑤
                }).done(data => $("#message").text(
                        data ? "注册成功!" : "该用户名已被注册!")                  ⑥
                ).fail(() => $("#message").text("无法提交注册!"));
            }
        );
    });
</script>
```

说明：

① 为了安全起见，对注册用户的密码使用 MD5 加密后再写入数据库。MD5 是一种不可逆、非对称加密算法，感兴趣的读者可查阅相关资料。为了实现 MD5 加密，这里引入了谷歌公司开发的基于 JavaScript 的 CryptoJS 库。CryptoJS 库支持 MD5、AES、DES 等多种加密算法。若要使用 MD5 加密，需要下载 core.min.js、lib-typedarrays.min.js 和 md5.min.js 这三个文件，大家可以去 https://www.bootcdn.cn/crypto-js 查看、下载相关资源。

② 以 POST 方式调用后台的控制器方法。

③ 由于后台使用@RequestBody 获取请求体中的 JSON 数据，所以这里必须将待注册用户的数据，转化成 JSON 字符串后才能发送。

④ 将密码用 CryptoJS.MD5()函数进行加密，并转换成字符串。

⑤ 设置向后台发送数据的格式必须是 JSON 字符串。

⑥ 接收后台 usr/regist 返回的数据 data，这个 data 显然是布尔型值。

3）后端：修改 Users 实体类

Users 类代码的改动很少，只是加了两个注解（粗体代码）：

```
@Entity
public class Users implements Serializable {
    @Id
    private String username;
    private String password;
    …
}
```

4）后端：UsersRepository 接口类

```
public interface UsersRepository extends JpaRepository<Users, String> {
    Users getByUsername(String username);
}
```

接口类中的 getByUsername() 方法,正是使用了表 9-2 中的关键词命名规范。这个方法的目的,就是注册时先查找数据库中该用户名是否已被注册过。整个接口类很简单,在此不做过多解释。

5) 后端:UsersDao 数据访问处理类

```
@Repository
public class UsersDao {
    private UsersRepository uRepository;

    @Autowired
    public UsersDao(UsersRepository uRepository) {                    ①
        this.uRepository = uRepository;
    }
    public boolean addUser(Users user) {    //返回布尔型值,跟前端呼应
        boolean tag = false;
        if (uRepository.getByUsername(user.getUsername()) == null) {  ②
            uRepository.saveAndFlush(user);                           ③
            tag = true;
        }
        return tag;
    }
}
```

说明:

① 通过 @Autowired 自动装配方式,Spring Boot 就会自动构建出 UsersRepository 对象 uRepository,并将其赋值给类变量 uRepository。这里采用的是通过 @Autowired 注解构造方法,来实现自动注入。

② 查找该用户名是否已被注册过,若没被注册(null 值),则执行③。大家可以自己做个试验:如果去掉这条语句,用已经注册过的用户名注册但其他信息填写不一样(例如修改一下 E_mail),提交注册后,数据库里面的数据有无变化,来进一步理解表 9-2 中的说明。

③ 将 user 持久化到数据库。

注意:为简化问题,这个方法并没考虑注册出错的情况。

6) 后端:UsersService 业务服务类

```
@Service
public class UsersService {
    private UsersDao userDao;
    public UsersService(@Autowired UsersDao userDao) {      //自动注入 UsersDao
        this.userDao = userDao;
    }
    public boolean regist(Users user) {
        return userDao.addUser(user);
    }
}
```

7）后端：UsersRegist 控制器类

```
@RestController
@RequestMapping("usr")
public class UsersRegist {
    private UsersService usersService;
    @Autowired
    public UsersRegist(UsersService usersService) {
        this.usersService = usersService;
    }
    @PostMapping("regist")     //访问路径对应 jQuery 代码中的"usr/regist"
    public ResponseEntity doLogin(@RequestBody Users user) {
        return ResponseEntity.ok().body(usersService.regist(user));
    }
}
```

ResponseEntity 是 Spring Boot 封装的、基于 HttpEntity 的常用于向前端传送数据的响应实体。可通过 ResponseEntity 向前端传送简单数据类型（例如字符串）：

```
return ResponseEntity.ok().body("登录成功!");
```

也可以传送 Java 对象，例如一个包含很多 Student 对象的列表：

```
return ResponseEntity.ok().body(list);
```

或者，包含数据对的 Map 对象：

```
Map<String, String> map = new HashMap(3);
map.put("username", "tom");
map.put("role","student");
map.put("email", "tom007@126.com");
return ResponseEntity.ok().body(map);
```

也可以向前端返回 HTTP 状态代码表示出错，例如：

```
return ResponseEntity.status(HttpStatus.UNAUTHORIZED).build();
```

当然，ResponseEntity 也可以设置 HTTP 响应头：

```
HttpHeaders headers = new HttpHeaders();
headers.add("Authorization", "success!");
return ResponseEntity.ok().headers(headers).body(map);
```

上面的代码，传送 map 数据的同时，通过响应头附加了额外信息。

9.3 Reactive 响应式处理

9.3.1 响应式概述

1. 传统编程缺点

传统的阻塞式编程容易造成资源的争用、浪费或并发问题，这时常通过增加硬件资源、多线程并行化来提高效率，但并行化方法不是万能的，仍然还会存在较多问题。在 Java Web 实践中，人们开始通过使用异步来处理任务，在异步处理完成后通过回调函数返回到当前进程继续处理，以便提高效率、解决问题。不过，如果回调函数太多，会导致代码的可读性、易维护性大幅降低。

Reactive 响应式编程旨在解决 Java Web "经典"异步方法的缺点。

2. Reactive 响应式

Reactive 响应式是指数据状态变化时能够及时做出响应而非被阻塞（阻塞意味着必须等待数据结果）。Reactive 是基于观察者设计模式的扩展而出现的，是一种异步编程范式。在 Reactive 中，数据（或事件）作为数据流，被发布和订阅，而异步任务则被组合、编排、运行或重用。程序员更多地关注于表述需要处理的问题（业务逻辑），而无须重点关注如何控制过程，例如只需要表述有两个数据流：看电影数据流、吃饭数据流，至于是先看电影还是先吃饭，或者是边吃边看，响应式代码提供了丰富的组合选项，形成链式数据流，类似于搭建好的"流水线"自动传递状态变化并做出响应。

3. Reactor

Reactor 是与 Spring 密切合作开发的、响应式编程范例的良好实现，需要 JDK8 以上版本才能使用。Reactor 提供了丰富的响应式操作要素，是后面要学习、使用的 Spring WebFlux 的首选响应库。Reactor 也为 Java Web 提供了良好支持。

尽管响应式编程并不一定会使应用程序运行得更快，但可以提高 Web 应用的弹性和吞吐量，有利于构建有韧性、弹性和消息驱动的异步、非阻塞式企业级应用系统。未来，响应式技术必将在 Java Web 应用开发中大显身手！

9.3.2 Reactor 基本原理

可以用打比方的方式来理解 Reactor 基本原理：

一位主编、五位作者（Publisher，发布者）和一家出版社（Subscriber，订阅者）合作完成一本书的编写。主编提交拟定纲要，发布出去，成为最初的发布者。每位作者提交写好的章节，提交给发布者，形成新的发布者实例。如此下去，形成一个基于异步流程的数据链。数据从第一个发布者发布的纲要开始，状态不断变化，最终订阅者编辑出版完成整个过程。

当然，如果没有出版社（订阅者）去向作者们（发布者）约稿（订阅），这意味着不会发生任何事情。

9.3.3 Reactor 核心包 publisher

Reactor 有一个至关重要的核心包 reactor.core.publisher，提供了绝大部分响应式操作的接口或实现，主要分为以下三大类。

◇ Mono：具备基本响应操作能力的流发布器，可异步发射 0 或 1 个项目。
◇ Flux：具备响应操作能力的流发布器，可异步发射 0 个或 N 个项目。
◇ Sinks：一种可以使用 Flux 或 Mono 推送响应流信号的结构，类似于接收器(或者汇总器)。

9.3.4 单量 Mono < T >

Mono < T >是 Reactor 的主要类型之一，是一种单量，这意味着 Mono 不会限制其发布的数据是什么类型。Mono 可以是一种发射数据后无须回应模式，也可以是一种"请求-响应"模式，还可将多个 Mono 连接在一起产生一个 Flux，非常灵活。下面，通过一些示例来学习其用法。

```
Mono < String > empty = Mono.empty();              //空量
Mono < String > wuhan = Mono.just("武汉");          //创建字符串量
Mono < Integer > num = Mono.just(100);
Mono < String > postcode = Mono.just("027");
empty.subscribe(System.out::println);              //订阅：输出 Mono 中的数据
wuhan.subscribe(System.out::println);
num.subscribe(System.out::println);
```

9.3.5 通量 Flux < T >

Flux < T >是 Reactor 的另一个主要类型，是一种通量。Flux 将多个数据项作为一个数据流(流量)来处理。可以将 Flux < T >转换为一个 Mono < T >，也可以将多个 Mono < T >连接合并成一个 Flux < T >。下面看一些示例。

示例 1：合并两个 Mono 并输出其内容。

```
wuhan.concatWith(postcode).subscribe(System.out::print);//输出：武汉 027
```

示例 2：将 1~100 中的偶数除以 2、奇数乘以 2 后输出。

```
Flux.range(1, 100)
    .map(i -> {
        if (i % 2 == 0) return i / 2;
        else return i * 2;
    }).subscribe(System.out::println);
```

map 是常用的操作符，常用于将数据流转换成映射表，然后就可以对映射表中的元素进行需要的处理。这里将 Flux 转换成映射表，然后对映射表中的每个元素进行除以 2 或乘以 2 处理。

示例 3：输出各个学院名称。

```
List < String > list = Arrays.asList("em 管理学院", "ac 会计学院", "mc 传媒学院", "ec 经济学院");
Flux < String > schools = Flux.fromIterable(list);        ①
schools.flatMap(s -> {                                     ②
    System.out.println(s);
    return Mono.empty();
}).subscribe();
```

说明：

① 除了可利用 just() 方法创建 Flux，还可以用 fromIterable() 方法。

② flatMap 是另外一种常用的流操作。flatMap 能够将每个 Flux 元素转换或映射为单个数据流，这样就可以对每个流进行按需处理，最后又将每个流合并为一个大的数据流返回。这里通过 flatMap 转换为单个流后，输出流中的每个元素 s。最后，由于是演示性质，并不需要合并流返回，所以这里返回了一个 Mono.empty() 空数据流。也可以用一句代码输出：schools.subscribe(s -> System.out.println(s))。

示例 4：向学院名称中临时增加一个学院"cc 数计学院"，以"序号 学院名称"方式输出全部学院名称。

```
List<String> list = Arrays.asList("em 管理学院", "ac 会计学院", "mc 传媒学院", "ec 经济学院");
Mono<String> news = Mono.just("cc 数计学院");
Flux<String> schools = Flux
        .fromIterable(list)                              //从集合中返回流
        .concatWith(news)                                //合并流
        .sort()                                          //对流进行排序
        .zipWith(Flux.range(1, list.size() + 1),         //生成 5 个数字,合并流
            (string, index) -> String.format(" %2d. %s", index, string));
                                                         //格式化输出
schools.subscribe(System.out::println);
```

concatWith 操作将一个流与另外一个流连接起来，有点类似于 Java 中的字符串连接。zipWith 操作则将当前流中的元素与另外流中的元素按照一对一的方式进行合并，也就是说将 5 个学院名称流与 5 个数字流，一对一合并。

控制台输出结果如下：

```
 1. ac 会计学院
 2. cc 数计学院
 3. ec 经济学院
 4. em 管理学院
 5. mc 传媒学院
```

示例 5：模拟不同时点会议的执行。

```
Mono<String> meeting1 = Mono.just("会议 1")
        .delayElement(Duration.ofMillis(2000));
Flux<String> meetings = Flux.just("会议 2", "会议 3", "会议 4")
        .delaySubscription(Duration.ofMillis(500))    //延迟 0.5s 发布流
        .delayElements(Duration.ofMillis(1000));      //延迟 1s 执行
Flux<String> merges = meeting1.mergeWith(meetings);   //合并
merges.map(s -> s + " 开始时间: " + System.currentTimeMillis())
        .subscribe(System.out::println);
Thread.sleep(6000);                                   //暂停以等待数据流结果,演示目的
```

执行后输出结果如下。

会议 2 开始时间：1629005389125
会议 1 开始时间：1629005389608
会议 3 开始时间：1629005390135
会议 4 开始时间：1629005391148

示例 6：过滤输出中国城市名称，并进行格式输出处理。

```
Flux.fromIterable(
        Arrays.asList("中国武汉","中国杭州","美国纽约","中国苏州","英国伦敦"))
    .transform(t -> t.filter(name -> name.contains("中国"))              ①
        .map(city -> city + "!"))
    .doOnNext(chinaCity -> System.out.println("我爱："  + chinaCity))    ②
    .subscribe();
```

说明：

① transform()是流转换操作。filter 操作则对流元素进行过滤。这里先利用 filter()过滤出 Flux 中每一项包含有"中国"的城市名称，然后在每一项数据后加"!"，完成转换操作。

② Flux 每发射一项数据就会调用一次 doOnNext()，所以在这里附加"我爱："信息比较恰当。运行后控制台输出如下形式数据。

```
我爱：中国武汉!
我爱：中国杭州!
我爱：中国苏州!
```

9.3.6 并行 ParallelFlux

在多核 CPU 大行其道的今天，Reactor 提供了 ParallelFlux 概念以便并行化操作。通过使用 parallel()方法，就可获得 ParallelFlux。这时候，数据流中的工作负载被划分为若干个能够并行运行的"轨道"，每个"轨道"可执行一个操作。轨道的数量，默认是与 CPU 内核数相同。

示例：并行输出学院名称。

```
List<String> list = Arrays.asList("em 管理学院","ac 会计学院","mc 传媒学院","ec 经济学院");
Flux.fromIterable(list).parallel(Schedulers.DEFAULT_POOL_SIZE)           ①
        .runOn(Schedulers.parallel())                                     ②
        .map(s -> s + " 线程名：" + Thread.currentThread().getName() + " 时间：" + System.currentTimeMillis())
        .subscribe(System.out::println);
```

说明：

① 该语句用于启用 ParallelFlux，轨道数量与 CPU 内核数相同。

② 运行于并行运行轨道上，Schedulers.parallel()并行处理调度器负责调度轨道操作。

在有四核 CPU 的计算机上测试上述代码，输出结果类似于下面这样，不同次测试的输出结果自然是不一样的。

```
em 管理学院 线程名: parallel-1 时间: 1629175160921
ec 经济学院 线程名: parallel-4 时间: 1629175160921
mc 传媒学院 线程名: parallel-3 时间: 1629175160921
ac 会计学院 线程名: parallel-2 时间: 1629175160921
```

9.3.7 处理槽 Sinks

Sinks 是 Reactor 提供的能以手工方式触发流数据信号的类。通过信号的触发,可以进行诸如推送新数据、重播旧数据等操作。表9-3 列出了 Sinks 的一些常用方法。

表 9-3 Sinks 常用方法

方法	说明
many().multicast()	多播。只向多个订阅者传输新推送数据,如果没有订阅者则接收的消息被丢弃
many().unicast()	单播。只向单个订阅者传输新推送数据,如果没有订阅者则保存消息,直到第一个订阅者订阅
many().replay()	重播。推送历史数据,并继续实时推送新数据
one()	向其订阅者播放信号(完成/出错),返回一个 Mono
empty()	向其订阅者播放完成或出错的终端信号,返回一个 Mono
tryEmitNext	异步发射下一个元素
asFlux()	转为 Flux
latest()	最新的一个元素

示例: 将消息 Note 推送给订阅者。

```
@Bean
public Sinks.Many<Note> broadcast() {
    return Sinks.many().replay().latest();
}
```

9.3.8 响应式 R2dbcRepository

前面利用 JpaRepository 实现了用户注册,Spring Boot 还提供了 R2dbcRepository 以进行响应式数据访问。使用方式与 JpaRepository 类似,先自定义接口类扩展自 R2dbcRepository。然后在自定义接口中编写各种数据访问方法。

除了可继续使用 JPQL 关键词范式外,还可以使用如表9-4 所示的一些常用方法。

表 9-4 R2dbcRepository 常用方法

方法	说明
Flux<T> findAll()	查询全部 T 对象
Flux<T> findAllById(Iterable<ID> var1)	根据指定的关键字 var1 集合进行查询
Mono<T> findById(ID var1)	根据指定的关键字 var1 查询 T 对象
Mono<S> findOne(Example<S> var1)	利用 Example 范例模式查询 S 型对象 var1

续表

方　法	说　明
Flux＜S＞ findAll(Example＜S＞ var1)	利用 Example 范例模式查询全部 S 型对象 var1
Mono＜Boolean＞ existsById(ID var1)	是否存在指定关键字 var1 的对象
Mono＜S＞ save(S var1)	保存 S 型对象 var1
Flux＜S＞ saveAll(Iterable＜S＞ var1)	保存集合 var1 中的 S 型对象
Mono＜Void＞ deleteById(ID var1)	根据指定关键字 var1 删除对象
Mono＜Void＞ delete(T var1)	删除 T 型对象 var1
Mono＜Void＞ deleteAll(Iterable＜? extends T＞ var1)	删除集合 var1 中的对象
Mono＜Void＞ deleteAll()	删除全部对象

示例：将用户数据保存到数据库中。

```
public Mono<String> addUser(Users u) {
    return uRepository.findById(u.getUsername())     //根据用户名查找是否已存在
            .flatMap(s -> Mono.just("existed"))      //数据表中已存在该记录
            .switchIfEmpty(                          //如果未查询到,说明未注册过
                    uRepository.registUser(          //写入数据库中
                            u.getUsername(), u.getPassword(), u.getLogo(),
                            u.getRole(), u.getEmail())
                            .map(s -> "sucess"))
            .onErrorResume(e -> Mono.just(HttpStatus.EXPECTATION_FAILED.toString()));
}
```

.onErrorResume()返回出错状态码 417 给前端。上述代码中 uRepository 对应的接口类如下：

```
public interface UsersRepository extends R2dbcRepository<Users, String> {
    @Modifying
    @Query("insert into users values(:username,:password,:logo,:role,:email)")
    Mono<Boolean> registUser(String username, String password, String logo, String role, String email);
}
```

UsersRepository 接口类在后面的场景任务示例中还会用到。这个示例，其实就是将上一节中的用户注册改成了 R2dbcRepository 实现。不过,这里并没有提供全部修改后的代码,剩余代码交给读者自行完成!

提示：本章及后续章节将使用 R2dbcRepository,不再使用 JpaRepository。

9.3.9　启用响应式 R2DBC

1. 简介

R2DBC 是响应式关系数据库连接 Reactive Relational Database Connectivity 的缩写。

前面使用的 JDBC 是一个完全阻塞的 API,即使用 HikariCP 连接池补偿阻塞行为的效果也是有局限的,而 R2DBC 声明了反应式 API,用异步、非阻塞方式来处理访问数据库时的并发性,能够充分利用多核 CPU 的硬件资源处理并发请求,利用较少的硬件资源就能够进行扩展。

R2DBC 需要 JDK8 以上版本支持。

2. 加入依赖

在 pom.xml 中加入 R2DBC 的依赖：r2dbc-postgresql 驱动和 spring-boot-starter-data-r2dbc。详细代码请参阅稍后的 pom.xml 依赖资源盘点情况。

3. 配置 r2dbc 连接属性

先清空 application.properties 文件中的原有内容，配置以下代码：

```
spring.r2dbc.url = r2dbc:postgresql://localhost:5432/tamsdb
spring.r2dbc.username = admin
spring.r2dbc.password = 007
spring.r2dbc.pool.enabled = true
spring.r2dbc.pool.initial-size = 5
spring.r2dbc.pool.max-size = 20
spring.r2dbc.pool.max-idle-time = 15000//25 分钟
```

为了避免出现依赖资源混乱的状况，下面盘点一下 pom.xml 中都包含有哪些依赖资源：

```xml
<dependencies>
    <dependency>
        <groupId>io.r2dbc</groupId>
        <artifactId>r2dbc-postgresql</artifactId>
        <version>0.8.8.RELEASE</version>
    </dependency>
    <dependency>
        <groupId>org.springframework.boot</groupId>
        <artifactId>spring-boot-starter-web</artifactId>
    </dependency>
    <dependency>
        <groupId>org.springframework.boot</groupId>
        <artifactId>spring-boot-starter-data-r2dbc</artifactId>
    </dependency>
    <dependency>
        <groupId>org.springframework.boot</groupId>
        <artifactId>spring-boot-starter-test</artifactId>
        <scope>test</scope>
    </dependency>
</dependencies>
```

9.3.10 启用分布式内存网格

1. Hazelcast 简介

考虑这样一种场景：某企业的 Web 应用系统分布、配置在若干台相互独立却又高速互联的不同地域的计算机上（集群）。当成千上万的用户查询某特定产品数据时，其实用户查的是同样的数据，做的是重复性工作，可是却需要不断去连接数据库获取数据。那么，能否将某位用户（例如第一位用户)从某台计算机(节点)数据库中查询到的数据,保存到集群中的其他计算机的内存里?这样其他用户查询时,直接从集群中不同节点内存中读取数据就可以了,速度快很多!再比如:一个用户从某个节点登录成功后,需要到其他不同节点获取信息,那么其登录状态能否也分享到集群中其他节点?其他节点只需要检查其登录状态是否有效即可,方便快捷很多!

Hazelcast 是一个开源分布式数据分发、集群的内存网格化工具,可以帮助解决上述问题。Hazelcast 完全基于内存来存储数据,可以实现共享各种数据状态、数据集合以及其他各类信息,并能实现数据在不同节点计算机之间的自动平衡。轻量化、简单易用、高可扩展性、高可用性(官方宣传 100% 可用、从不失败),能够实现基于云传输的高速数据共享处理,可以和任何现有数据库系统一起使用。Hazelcast 在全球拥有广泛用户,例如 Vert. x、Pentaho、MuleSoft、Apache Drill 和 Accor 等。

2. 添加 Hazelcast 依赖

在 pom.xml 中加入:

```xml
<dependency>
    <groupId>com.hazelcast</groupId>
    <artifactId>hazelcast-all</artifactId>
    <version>4.2.2</version>
</dependency>
```

3. 修改入口类

```java
@EnableCaching
@SpringBootApplication
public class Chapter9Application extends SpringBootServletInitializer {
    ...
}
```

加粗代码为新加的。@EnableCaching 启用缓存。Spring Boot 会自动将 Hazelcast 作为缓存管理器。

4. 使用配置文件

在项目的 resources/static 下新建文件夹 config。

展开项目的 External Libraries,再依次展开 Maven:com.hazelcast:hazelcast-all:4.2.2→hazelcast-all-4.2.2.jar,如图 9-5 所示。找到 hazelcast-default.xml,复制到项目的 resources/static/config 下,如图 9-6 所示。

打开 application.properties,设置 Hazelcast 配置文件的位置:

```
spring.hazelcast.config = classpath:static/config/hazelcast-default.xml
```

启动项目,控制台将输出 Hazelcast 配置信息,说明 Hazelcast 成功启用。图 9-7 是在两台 IP 地址分别为 192.168.1.5 和 192.168.0.102 的计算机上启用 Hazelcast 的运行情况。

在后面 9.5 节的场景应用示例中,将会使用 Hazelcast 缓存并分享数据。

图 9-5 hazelcast-default.xml　　　图 9-6 复制 hazelcast-default.xml

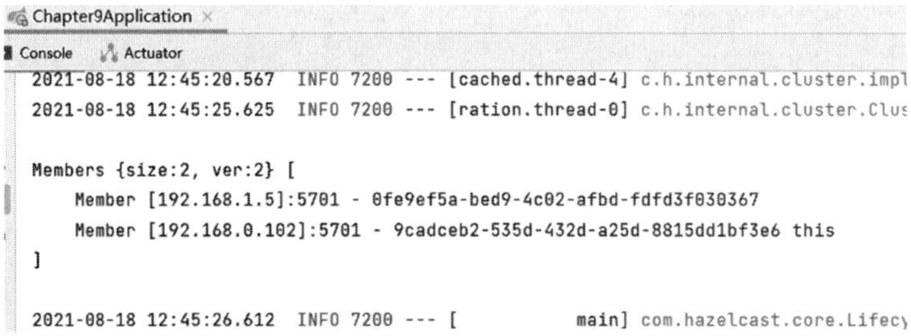

图 9-7 Hazelcast 成员节点

9.4 Spring WebFlux

9.4.1 Spring WebFlux 简介

Spring WebFlux(以下简称 WebFlux)是自 Spring 5.0 开始提供的异步、完全非阻塞式的响应式 Web 框架,能够充分利用多核 CPU 的硬件资源去处理大量请求任务。WebFlux 内部使用的是前面学习过的响应式编程,并以 Reactor 响应库为基础,可以在不扩充硬件资源的情况下,提升系统的吞吐量和伸缩性。

WebFlux 依赖于非阻塞 I/O 进行任务处理,因此在 Tomcat、Jetty、Servlet3.1 以上及 Netty、Undertow 等非 Servlet 应用服务器上,都可以运行。

本章创建的项目是基于 Spring MVC 的,而 Spring MVC 依赖于阻塞式 I/O。现在先学习好相关知识,打好基础,在第 13 章将用 WebFlux 完整实现一个案例项目。

9.4.2 WebFlux 应用的入口类

Spring Boot 提供了一个 WebFlux 启动器,实现了很多自动化配置工作。WebFlux 应用的入口类代码非常简单:

```java
@SpringBootApplication
public class TamsApplication {
    public static void main(String[] args) {
        SpringApplication.run(TamsApplication.class, args);
    }
}
```

读者可与本章的 Chapter9Application 入口类进行比较,体会其差异。

9.4.3 配置 WebFlux 应用

1. 配置静态资源

默认情况下,放置在 META-INF/resources、/resources、/static、/public 下的静态资源(例如 HTML 文件、图片、CSS、JS 文件)可直接访问,无须做任何配置。当然,也可显式地对这些资源进行配置:

```java
@Configuration
@EnableWebFlux
public class WebConfig implements WebFluxConfigurer {
    @Override
    public void addResourceHandlers(ResourceHandlerRegistry registry) {
        registry.addResourceHandler("/resources/**")
                .addResourceLocations("/public");
    }
}
```

@Configuration 注解该类为配置类,@EnableWebFlux 启用 WebFlux 配置。

2. 细节化配置

WebFlux 的默认配置基本能够满足应用要求,不过更常见的是进行一些自定义配置。可以定义一个类 WebConfig,让这个类实现 WebFlux 提供的 WebFluxConfigurer 接口,即可对应用进行细节化配置,例如:

```java
@Configuration
@EnableWebFlux
public class WebConfig implements WebFluxConfigurer {
    @Override
    public void addFormatters(FormatterRegistry registry) {
        DateTimeFormatterRegistrar reg = new DateTimeFormatterRegistrar();
        reg.setDateFormatter(DateTimeFormatter.ofPattern("yyyy-MM-dd HH:mm:ss"));
```

```
            reg.setDateStyle(FormatStyle.LONG);
            reg.registerFormatters(registry);
    }
}
```

这里改写了 WebFluxConfigurer 接口的 addFormatters()方法,以便对 Web 应用程序的日期格式进行自定义设置。

9.4.4 HandlerFilterFunction 操作过滤

有时,需要对用户的请求操作进行过滤,例如对已经登录失效的用户,过滤掉其请求并返回相应提示信息。或者,根据用户身份的不同,进行不同的操作处理。WebFlux 提供了 HandlerFilterFunction 接口,可以对请求操作进行过滤处理,只需实现该接口并重写 filter()方法,例如:

```
@Component
public class AuthFilter implements HandlerFilterFunction {
    @Override
    public Mono filter(ServerRequest request, HandlerFunction next) {
        String token = request.headers().firstHeader(HttpHeaders.AUTHORIZATION);
        if (verify(token))                                              //用户身份有效
            return next.handle(request);                                //继续处理
        else
            return ServerResponse.status(UNAUTHORIZED).build();         //非授权用户
    }
}
```

然后,在需要的地方,应用该过滤器即可。

9.4.5 HandlerFunction 业务处理

在 WebFlux 中,HTTP 请求可由 HandlerFunction 处理:这是一个接受 ServerRequest 并返回延迟的 ServerResponse 的函数接口。HandlerFunction 类似于前面学过的被@RequestMapping 注解的方法的主体。ServerRequest 为输入的请求,ServerResponse 为响应输出的数据。例如:

```
HandlerFunction<ServerResponse> hello = request ->
                    ServerResponse.ok().bodyValue("Hello,Spring WebFlux!");
```

不过,一旦功能处理较多,这种写法显得有些不好管理。更通行的做法是采用跟 Spring MVC 类似的代码组织方式,将处理同类业务的 HandlerFunction 组织在一个类中,例如:

```
@Component
public class UsersHandler {
    public Mono<ServerResponse> login(ServerRequest request) {
        ……
    }
```

```
    public Mono<ServerResponse> regist(ServerRequest request) {
        ……
    }
}
```

9.4.6　RouterFunction 路由函数

在前面都是使用@RequestMapping()、@PostMapping()和@GetMapping()等对用户请求地址进行映射，WebFlux 可通过 RouterFunction 接口进行函数式路由，这样更方便管理路由。例如上面的 hello 函数：

```
RouterFunction<ServerResponse> router = RouterFunctions.route(GET("/sayhello"), hello);
```

同样，更好的组织方式是将同类路由地址放在一起：

```
@Configuration
public class RouteConfig {
    @Bean
    public RouterFunction<ServerResponse> usrRoute(UsersHandler handler) {
        return route()
                    .path("/usr", r -> r                               ①
                                .POST("/login", handler::login)        ②
                                .POST("/regist", handler::regist))
                    .build();
    }
    …
}
```

说明：
① 设置该业务处理的根路径为/usr，下面再进行业务处理路径的分支管理。
② 设置该业务处理的路径为/login，并调用 UsersHandler 类的 login 方法进行处理。假定站点地址为 http://localhost，则用户登录访问地址为 http://localhost/usr/login。类似地，用户注册访问地址为 http://localhost/usr/regist。

这里使用@Bean 注解 usrRoute()为一个功能处理单元。

9.4.7　Multipart Data 多域数据

在第 5 章 5.8.1 节的文件上传示例中，在表单中设置了 enctype="multipart/form-data"来提交混合型数据：既包含文本又包含文件。WebFlux 提供了 MultipartBodyBuilder 类来构建 Multipart 数据。

示例：模拟用户注册时，用户名、用户头像的提交。

```
MultipartBodyBuilder builder = new MultipartBodyBuilder();
builder.part("username", "杨过");                                    //用户文本数据
```

```
Resource image = new ClassPathResource("static/image/me.png");     //用户头像文件
builder.part("logo", image);
MultiValueMap<String, HttpEntity<?>> formData = builder.build();
```

上述代码构建了一个完整的表单提交数据体,现在可将该数据体提交给后端的处理程序:

```
WebClient.builder().baseUrl("/file/up").build()
        .post().body(BodyInserters.fromMultipartData(formData))
        .retrieve().toBodilessEntity().subscribe();
```

包含用户名、用户头像文件的数据体被提交给映射地址为 file/up 的某个后端处理程序。现在,就可以在后端处理程序中接收用户名、用户头像文件:

```
request.multipartData().flatMap(m -> {
        Map<String, Part> parts = m.toSingleValueMap();
        String username = ((FormFieldPart) parts.get("username")).value();
        FilePart part = (FilePart) parts.get("logo");
        … //其他后续处理
});
```

这里的 request 当然是服务请求类型 ServerRequest。FormFieldPart 表示表单的文本域,FilePart 自然是表单的文件域。如果只需要处理多域数据中的文件数据,还可以专门从多域数据中抽取出文件:

```
public Mono<ServerResponse> processFile(ServerRequest request) {
    return request.body(BodyExtractors.toParts())
            .filter(part -> part instanceof FilePart)      //过滤出文件域
            .next()
            .cast(FilePart.class)
            .flatMap(part -> {
                    String fileName = part.filename();
                    …
            });
}
```

9.5 场景应用示例

9.5.1 学生信息查询

在第 8 章中用 Servlet 方式实现了学生信息查询。读者对架构、实现过程,应该还有印象。这次,我们改用 Spring Boot 来实现,继续沿用第 8 章的思路。

1. 前端:HTML 页面 query.html

无须改变。

2. 前端：query.js

改动很少，只需要将语句：

```
$.getJSON("query.do", {sname: $("#sname").val()})
```

修改为：

```
$.getJSON("stu/query/" + $("#sname").val())
```

这里使用了 RESTful 风格的请求地址。

3. 后端：Student 实体类

原有的写法有些复杂，为了简化语句的编写，我们使用第三方 lombok 插件项目。首先，在 pom.xml 中加入 lombok 依赖：

```xml
<dependency>
    <groupId>org.projectlombok</groupId>
    <artifactId>lombok</artifactId>
    <version>1.18.20</version>
    <scope>provided</scope>
</dependency>
```

现在，Student 实体类代码非常简洁，变成这样：

```java
@Data
@Table
@NoArgsConstructor   //生成无参构造方法
@AllArgsConstructor  //生成全参构造方法,即含全部属性的构造方法
public class Student implements Serializable {
    @Id
    private String sno;
    private String sname;
    private String sclass;
    private String tel;
    private String address;
    private String postcode;
}
```

4. 后端：StudentRepository 数据访问接口类

这个类第 8 章并没有，属于新建类。代码如下：

```java
public interface StudentRepository extends R2dbcRepository<Student, String> {
    Flux<Student> findStudentsBySname(String sname);
}
```

利用 JPQL 关键词范式,查询数据库并返回 Flux 结果。

5. 后端:StudentDao 业务逻辑处理类

```
@Repository
@RequiredArgsConstructor
public class StudentDao {
    private final @NonNull StudentRepository sRepository;        //由 lombok 帮助注入
    public Flux<Student> findStudentsBySname(String sname) {
        return sRepository.findStudentsBySname(sname);
    }
}
```

代码非常清晰易懂。

6. 后端:StudentService 服务类

```
@Service
@RequiredArgsConstructor
public class StudentService {
    private final @NonNull StudentDao studentDao;
    public Flux<Student> queryStudentsByName(String sname) {
        return studentDao.findStudentsBySname(sname);
    }
}
```

7. 后端:StudentController 控制器类

```
@RestController
@RequestMapping("stu")
@RequiredArgsConstructor
public class StudentController {
    private final @NonNull StudentService studentService;
    @GetMapping("query/{sname}")                                //访问路径类似于:stu/query/杨过
    public List<Student> queryStudentsByName(@PathVariable String sname) {
        return studentService.queryStudentsByName(sname)
                .toStream()
                .sorted(Comparator.comparing(Student::getSno).reversed())    ①
                .collect(Collectors.toList());      //转换成 List 数据
    }
}
```

说明:

① 这里对流中的数据进行了排序:按照学号降序排列,这样同一个专业的学生就可以显示在一起,便于查看。

9.5.2 基于 JWT 令牌实现分布式登录

视频讲解

1. 应用需求

应用需求并不复杂,注意以下三点。

(1) 用户登录成功后,生成 JWT 令牌,以便后续其他场景使用;
(2) 以 Mono＜ResponseEntity＞流数据的形式,通过 HTTP 头 Authorization 挂载令牌信息;
(3) 使用两台计算机,模拟集群中的两个节点:在一个节点成功登录后,再去另外一个节点登录时,会提示已经登录过。

2. 处理思路

生成 JWT 令牌,需要熟悉 JWT 相关知识,这里使用 jboss.resteasy 来生成令牌。至于模拟集群,自然是使用前面介绍的 Hazelcast 了。整个处理架构如图 9-8 所示。

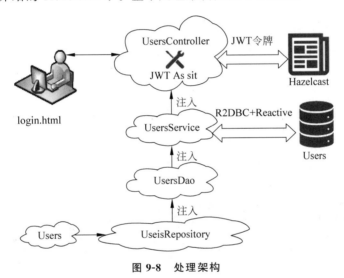

图 9-8 处理架构

3. JWT 令牌简介

设想某 Web 系统放置在 A、B 两个服务器上,当用户从 A 登录成功后,传统的处理方法是用 session 保存用户登录状态。但若用户需要去 B 下载授权资料,session 在 B 上就不能发挥作用了,JWT 令牌则可以解决类似这样的问题。

JWT 是 JSON Web Token 的简称,基于工业化标准 RFC7519,是一个开放的标准协议。JWT 能够以 JSON 格式安全地传输用户登录状态,也为跨域问题提供了支持。

JWT 由三个部分组成:Header、Payload 和 Signature。

1) Header 头部

描述关于该 JWT 的最基本信息:

```
{
    "alg": "HS256",
```

```
    "typ":"JWT"
}
```

alg定义了使用的算法。该JSON对象需要转换为经过Base64URL编码的字符串(数据①):

```
ewoJImFsZyI6ICJIUzI1NiIsCgkidHlwIjogIkpXVCIKfQ
```

2) Payload 载荷

JWT 的主体内容部分,也是一个 JSON 对象,包含需要传递的数据。JWT 指定了以下七个默认属性以供选择使用,并允许自定义属性。

◇ sub：主题。
◇ iss：签发人。
◇ iat(issued at)：签发时间。
◇ exp(expires)：失效时间。
◇ jti(JWT id)：标识 ID。
◇ aud：受众。
◇ nbf(not before)：在此时间前不可用。

这些属性并不是每个都必须定义。例如：

```
{
    "sub":"JWT",
    "iss":"http://www.hust.edu.cn",
    "iat":1451888119,
    "exp":1454516119,
    "jti":"hust0102038",
    "aud":"EM",
    "myname":"ccgg"
}
```

其中 myname 是自定义属性。Payload 也需要经过 Base64URL 进行编码,生成字符串(数据②):

```
ewogICAgInN1YiI6ICJKV1QiLAogICAgImlzcyI6ICJodHRwOi8vd3d3Lmh1c3QuZWR1LmNuIiwKIC
AgICJpYXQiOiAxNDUxODg4MTE5LAogICAgImV4cCI6IDE0NTQ1MTYxMTksCiAgICAianRpIjogImh1
c3QwMTAyMDM4IiwKICAgICJhdWQiOiAiRU0iLAogICAgIm15bmFtZSI6ICJjY2dnIgp9
```

3) Signature 签名

先将上面①、②两部分数据组合在一起,中间以"."隔开,即：①.②。然后,通过指定的密钥及算法,对其进行运算生成签名。例如,如果使用密钥007、SHA-256加密算法对①.②进行加密,则最后得到签名(数据③):

```
T5ucJ3Lzf9u3t5mdcdBTaNwS1UG4QVJL3HElOfEvnwA
```

将这三部分数据组合在一起,每个部分之间用"."分隔,即"①.②.③"的形式,组合成一个字符串,就构成整个 JWT 对象。

4. 使用 RestEasy 处理 JWT

RestEasy 是著名的 Red Hat 开源解决方案提供商的一个官方项目,用于构建 RESTful Web 服务和 Java 应用程序,是 Eclipse 基金会 Jakarta RESTful Web Services 规范的实现。RestEasy 可运行在任何 Servlet 容器中,具有丰富的特性。

RestEasy 提供了便捷的生成、加密、验证 JWT 的方法。利用 JsonWebToken 类可轻松生成 JWT,利用 JWSBuilder 类可对 JWT 进行加密,而 RSAProvider.verify 则可对 JWT 进行验证。

要使用 RestEasy 处理 JWT,需要在 pom.xml 中加入 RestEasy JWT 依赖:

```xml
<dependency>
    <groupId>org.jboss.resteasy</groupId>
    <artifactId>resteasy-jaxrs</artifactId>
    <version>3.15.1.Final</version>
</dependency>
<dependency>
    <groupId>org.jboss.resteasy</groupId>
    <artifactId>jose-jwt</artifactId>
    <version>3.15.1.Final</version>
</dependency>
```

5. 配置 Hazelcast

打开项目 resources/static/config/hazelcast-default.xml 文件。

1) 启用并添加节点计算机 IP 地址

```xml
<tcp-ip enabled="true">
    <member>192.168.1.5</member>
    <member>192.168.0.102</member>
</tcp-ip>
```

缓存的 JWT 令牌,将在这两台计算机之间共享。

2) 修改缓存 map

```xml
<map name="getToken">
    …
    <time-to-live-seconds>600</time-to-live-seconds>
    …
</map>
```

这里的 name,是指需要缓存数据的方法名。由于缓存的是 JWT 令牌,因此在后面的第 9 步骤,就会定义 getToken()方法。缓存时间可视需要进行设置,这里设置为 10 分钟。缓存到期后,再次请求时会重新去获取数据;否则,从缓存中获取数据。

6. 前端：login.html

```html
<link href="css/users.css" rel="stylesheet" type="text/css">
<script src="js/jquery-3.6.0.min.js"></script>
<script str="js/core.min.js"></script>
<script src="js/lib-typedarrays.min.js"></script>
<script src="js/md5.min.js"></script>
<script src="js/login.js"></script>
<script>
    $(document).ready(function () {
        $.doLogin();
    });
</script>
<div class="user-login">
    账户登录<br>
    <img src="image/username.png">
    <input type="text" class="login-input" placeholder="用户名" id="username">
    <br><br><img src="image/password.png">
    <input type="password" class="login-input" placeholder="密　码" id="password">
    <br><br><button class="login-button" id="loginBtn">登 录</button>
    <br><span id="message"></span>
</div>
```

7. 前端：login.js

```javascript
$.doLogin = function () {
    $("#loginBtn").click(function () {
        $.ajax({
            type: 'post',
            url: 'usr/login',                              //后端登录控制器地址
            data: JSON.stringify({                         //数据转换为JSON字符串
                username: $("#username").val(),
                password: String(CryptoJS.MD5($("#password").val()))    //MD5 加密
            }),
            contentType: "application/json"
        }).done((data, textStatus, jqXHR) => {
            if (data == "hasLogined") {                                                ①
                $("#message").text("已经登录过……");
                return
            }
            let auth = jqXHR.getResponseHeader("Authorization");                       ②
            if (auth == null)
                $("#message").text("用户名或密码错误,登录失败!");
            else
                $("#message").text("登录成功!");
        }).fail(() => $("#message").text("无法提交登录!"));
    });
}
```

说明：

① 如果端返回的数据是 hasLogined，则表示已在别的节点计算机登录过了，且登录未失效，无须重新登录。

② 获取后端响应返回的 HTTP 头 Authorization 属性的数据，据此来判断用户登录成功与否。相应地，一旦登录成功，后端代码会在返回给前端的响应数据中，设置 HTTP 头属性 Authorization 的具体值。

8. 后端：Users 实体类

```java
@Data
@Table
@NoArgsConstructor
@AllArgsConstructor
public class Users implements Serializable {
    @Id
    private String username;
    private String password;
    private String logo;
    private String role;
    private String email;
}
```

9. 后端：JWTAssit 令牌工具类

```java
@Component
public class JWTAssit {
    public static KeyPair KEY_PAIR;
    static {
        try {
            KEY_PAIR = KeyPairGenerator.getInstance("RSA").generateKeyPair();    ①
        } catch (NoSuchAlgorithmException e) {
            e.printStackTrace();
        }
    }
    @Cacheable(value = "getToken", key = "#username", unless = "#result == null")    ②
    public String getToken(String username) {
        String token = null;
        try {
            JsonWebToken jwt = new JsonWebToken();
            jwt.id(String.valueOf(new Random().nextLong()))                    //随机令牌编号
                .audience(URLEncoder.encode(username, "utf-8"))                ③
                .expiration(System.currentTimeMillis() + 60000 * 20)           //有效时长20min
                .issuer("ccgg")                                                //签发人
                .principal("myweb");                                           //主题
            token = new JWSBuilder()
                .contentType("text/plain")
                .jsonContent(jwt)
                .rsa256(KEY_PAIR.getPrivate());                                //生成令牌
```

```
        } catch (Exception e) { e.printStackTrace(); }
        return token;
    }
}
```

说明：

① 生成密钥对对象。通过该对象，可获取随机的私钥、公钥，以便加密、解密时使用。

② 缓存 getToken()方法返回的 JWT 令牌。value="getToken"，定义了缓存名称为 getToken，这个与前面 hazelcast-default.xml 中 name="getToken"是一致的。key 定义了缓存键名，这里使用登录用户名。unless 设置不进行缓存的条件：令牌为空时，意味着用户登录失败，不缓存数据。

③ 设置令牌的受让人，也就是令牌颁发给登录用户。使用 URLEncoder 是为了避免 token 中的中文用户名乱码问题。

10. 后端：UsersRepository 接口类

在前面的 9.3.8 节就已经创建好了，无须添加额外的代码。

11. 后端：UsersDao 数据访问处理类

```
@Repository
@RequiredArgsConstructor
public class UsersDao {
    private final @NonNull UsersRepository uRepository;
    public Mono<Users> findUser(Users user) {
        return uRepository.findOne(Example.of(user));
    }
}
```

Spring R2DBC 专门为高频查询业务提供了 Example 查询方式。只需要将需要查询的对象传递给 Example.of()方法后，就可通过 Example 适配器找到数据库中的对应记录，并通过 findOne()返回一个 Mono 对象。当然，完全可以通过 JPQL 关键词范式进行查询。

12. 后端：UsersService 业务服务类

```
@Service
@RequiredArgsConstructor
public class UsersService {
    private final @NonNull UsersDao userDao;
    public Mono<Users> login(Users user) {
        return userDao.findUser(user);
    }
}
```

13. 后端：UsersController 控制器类

```
@RestController
@RequestMapping("usr")
```

```
@RequiredArgsConstructor
public class UsersController {
    private final @NonNull UsersService usersService;
    private final @NonNull CacheManager cacheManager;          //缓存管理器
    private final @NonNull JWTAssit jwtAssit;                  //JWT令牌工具

    @PostMapping("login")
    public Mono<ResponseEntity> doLogin(@RequestBody Users user) {
        Cache tokenCache = cacheManager.getCache("getToken");  //获取getToken缓存
        if (tokenCache != null) {                              //缓存已存在
            Cache.ValueWrapper wrapper = tokenCache.get(user.getUsername());
            if (wrapper != null)                               //缓存中存在该用户名为键的令牌
                return Mono.just(ResponseEntity.ok().body("hasLogined"));
                                                               //已登录返回
        }
        return usersService.login(user)                        //缓存中无数据,则查询数据库
                .flatMap(u -> {                                //登录成功
                    Map<String, String> map = new HashMap(3);
                    map.put("username", u.getUsername());
                    map.put("logo", u.getLogo());
                    map.put("role", u.getRole());
                    HttpHeaders headers = new HttpHeaders();
                    headers.set("Authorization", jwtAssit.getToken(u.getUsername())); ①
                    return Mono.just(ResponseEntity.ok().headers(headers).body(map)); ②
                });
    }
}
```

说明:

① 这句代码做了两件事:一是 getToken() 方法生成 JWT 令牌,二是将该令牌附加在 HTTP 头属性 Authorization 中,最后通过 ResponseEntity 返回给前端。图 9-9 是某用户成功登录后在浏览器监测中显示的响应头 Authorization,Authorization 的内容即为生成的 JWT 令牌。

将 Hazelcast 缓存令牌的时间与令牌本身的有效期设置成一样(都是 10min),缓存失效即意味着令牌失效。Hazelcast 本身会进行缓存有效期的管理,但令牌是不是伪造的、是不是由合法用户颁发的或是否被人篡改等,Hazelcast 没办法进行检验,这里也没有使用 RestEasy 进行 JWT 有效性验证。关于 JWT 验证,将会在第 12 章介绍。

② 向前端返回两种数据:令牌,附加在 HTTP 响应头 Authorization 中;用户名、用户头像、用户角色,放在响应体中,以方便页面需要时使用。

需要注意的是,这里纯粹是演示、学习 Hazelcast 分布式缓存的使用思路,并不能作为实践中的登录处理方法,不然只要在另外一个节点计算机输入用户名,就能登录成功。不过,也有可以应用的场所,例如授权用户才能下载文件的处理。当用户登录成功后,可随机返回一个授权码凭证给用户,并由 Hazelcast 在节点之间缓存该凭证。只要在授权码凭证的有效期限内,输入该授权码凭证,即可在不同节点计算机下载所需的文件,无须再去访问数据库进行登录。

另外,9.5.1 节的学生信息查询,就比较适合使用缓存。这个任务,交给读者完成!

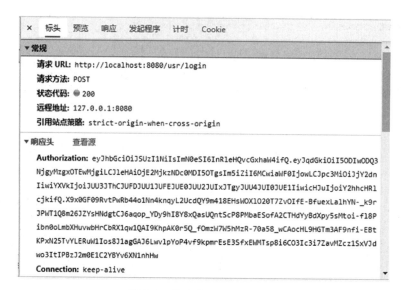

图 9-9 响应头 Authorization

9.6 场景任务挑战——模糊查询

在 9.5.1 节成功实现了学生信息查询。现在更改一下查询要求：根据输入的关键字，查询姓名或者班级中包含该关键字的全部学生记录，例如输入关键字“财”，则姓名中包含“财”字的(例如李章财)，或者班级中包含“财”字的(例如财管 11902)，都显示在页面上，如图 9-10 所示。

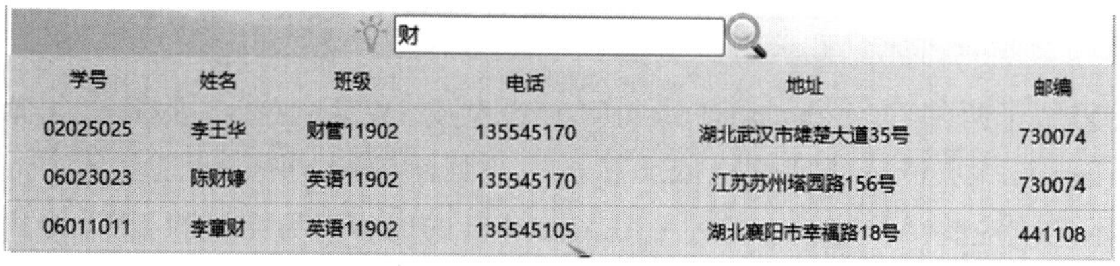

学号	姓名	班级	电话	地址	邮编
02025025	李王华	财管11902	135545170	湖北武汉市雄楚大道35号	730074
06023023	陈财婷	英语11902	135545170	江苏苏州塔园路156号	730074
06011011	李章财	英语11902	135545105	湖北襄阳市幸福路18号	441108

图 9-10 模糊查询

第10章 Vue.js渐进式框架

Vue.js 是一个渐进式 JavaScript 框架。本章主要介绍了 Vue 基本原理、基础语法、组合式和响应性函数、数据绑定、事件绑定及触发、自定义元素、渲染函数、组件、函数集、Axios 等内容。熟练掌握 Vue.js，将为开发基于数据驱动的 Web 视图提供强大支撑。

10.1 Vue 概述

Vue（官方拼读 /vjuː/）是一个专注于视图层数据展示的数据驱动、渐进式 JavaScript 框架。所谓数据驱动，就是只需要改变数据，Vue 就会自动渲染并展示新内容页面。渐进式的意思则是可以分阶段、有选择性地使用 Vue，无须使用其全部元素，小到输出"Hello,Vue!"的简单页面，大到复杂 Web 应用系统，增量使用。

Vue 易于学习，很容易与其他 JavaScript 库（例如 jQuery）集成使用，在实践中得到了广泛应用，受到业界极高的关注，被誉为 JavaScript 框架的三巨头之一。Vue 为高效率开发 Web 应用程序的前端页面提供了强大动力。

为更方便地使用 Vue，建议在 IntelliJ IDEA 中安装 Vue.js 插件，如图 10-1 所示。

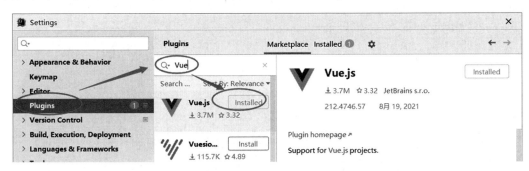

图 10-1 Vue.js 插件

10.2 Vue 应用基础

10.2.1 创建 Vue 应用

1. 安装

可以采用 NPM，或在页面引入 CDN 地址方式，例如：

```
<script src="https://unpkg.com/browse/vue@3.2.4/dist/"></script>
```

可以采取从官网下载 js 文件放入站点项目的方式。

vue.global.prod.js 是一个完整压缩版，将其复制到站点的 js 文件夹下，然后在页面引入即可：

```
<script src="js/vue.global.prod.js"></script>
```

2. 第一个 Vue 应用

我们来创建第一个 Vue 应用，在页面上输出"你好，Vue!"。为便于大家理解，这个示例给出了完整代码：

```
<!DOCTYPE html>
<html lang="zh">
<head>
    <meta charset="UTF-8">
    <title>hello Vue</title>
    <script src="js/vue.global.prod.js"></script>
</head>
<body>
<div id="app">                                    ①
    <span>{{welcome}}</span>
</div>
<script>
    const app = Vue.createApp({                   ②
        data() {                                  ③
            return {
                welcome: '你好,Vue!'
            }
        }
    })
    app.mount('#app')        //将 Vue 应用实例挂载到 id="app" 的层上面.
</script>
</body>
</html>
```

说明：

① 定义了一个层，用来挂载 Vue 应用的实例，通俗地说，就是让 Vue 来渲染这个层里面所有元素的数据及事件处理。打个比方就是，这个层类似于某个名叫"app"的公司，现在请 Vue 来接管该公司，对公司的各种数据业务（设备、资金、人事等）进行管理。所以，这个层里面标题<span>的数据，将由 Vue 渲染显示。

② 创建一个 Vue 应用的实例。Vue 会完成一系列初始化过程，例如设置数据、编译模板等，并运行一些被称为"生命周期钩子"的函数。利用这些钩子函数，就可以在适当的时候处理各种业务逻辑。这有点类似于乘客去乘车（id="app"的层），司机（Vue）要运行检查车辆、打火启动、起步环顾检查、发车等"钩子函数"，而乘客（层的内部元素）则可以在司机起步检查时处理自己的业务逻辑"调整坐姿"。

③ 通过 data() 函数，返回了一个 welcome 数据。当然，也可以返回多个数据。

还有一种创建 Vue 应用的方式：

```
<div id="app">
    <span>{{welcome}}</span>
</div>
<script>
    const MyData = {
        data() {
            return {
                welcome: '你好,Vue!'
            }
        }
    }
    Vue.createApp(MyData).mount('#app')
</script>
```

这种方式基于 MyData 根组件创建 Vue 应用，运行效果跟第一种方式完全相同。

10.2.2 生命周期

从创建 Vue 应用实例，到初始化、模板编译，再到渲染、挂载 DOM 节点，然后更新界面，最后销毁实例，完成了整个生成周期过程。读者可以去 Vue 官网参阅详细的生命周期流程图，以便更好地理解 Vue 的整个运行原理。

在整个生命周期过程中，Vue 提供了八个重要的钩子函数。下面选取了其中四个比较常用的钩子函数进行简要说明。

- ◇ created()：Vue 实例创建完成后调用。此时一些数据和函数已经创建好，但还没有挂载到页面上。可在这个函数里面进行一些初始化处理工作。
- ◇ mounted()：模板挂载到 Vue 实例后调用，HTML 页面渲染已经完成。可以在这个函数中开始业务逻辑处理。
- ◇ beforeUpdate()：页面更新之前被调用。此时数据状态值已经是最新的，但并没有更新到页面上。

◇ beforeDestroy()：解除组件绑定、事件监听等。在实例销毁之前调用。

示例：在待输出数据"你好，Vue!"后加上"渐进式框架!"字样。

```
const app = Vue.createApp({
    data() {
        return {
            welcome: '你好,Vue!'
        }
    },
    mounted() {
        this.addStr()
    },
    methods: {
        addStr: function () {
            this.welcome += '渐进式框架!'
        }
    }
})
app.mount('#app')
```

读者自行编写的若干函数，可以放置在 methods 里面。这里利用 mounted() 钩子函数，调用 addStr() 函数，将原有的数据进行了修改。打开页面后，将显示"你好，Vue! 渐进式框架!"。

注意：要使用数据或函数，需要用 this 进行限定，this 代表当前 Vue 实例。

10.2.3 组合式函数 setup()

为了更好地管理代码，Vue 提供了组合式函数 setup()。setup() 能够很好地将代码整合在一起，还可以有选择性地对外暴露我们定义的变量、常量和函数，基本语法格式如下：

```
const app = Vue.createApp({
    props: {
        addr: {type: String}
    },
    setup(props, context) {
        …
    }
})
```

◇ props：外部传入的数据，可以是数组或对象，这里传入的是一个 addr 字符串数据。后面结合数据绑定知识再举例。

◇ context：上下文对象，可用来获取应用的一些属性，context 和 props 都是可选参数。

示例：用 setup 函数，重写上一节的例子。

```
const app = Vue.createApp({
    setup() {
        const welcome = Vue.ref('你好,Vue!')
```

```
                const addStr = () => welcome.value += '渐进式框架!'
                Vue.onMounted(addStr)
                return {welcome}
            }
    })
    app.mount('#app')
```

较上一节写法,现在简洁明了很多。在 setup()函数里面,生命周期钩子函数的写法有所变化,原来的 mounted()变成了 onMounted(),类似的还有 onBeforeMount()、onBeforeUpdate()等。

Vue 是全局函数,里面定义了很多常量、函数或方法,例如 ref()、onMounted()等。我们在 Vue.onMounted()里面调用 addStr()函数,返回 welcome 对象添加内容后的值。

通过 setup()函数的 return,可以有选择性地对外暴露某些值或方法。这里如果不返回 welcome,<span>中是无法插值显示 welcome 的值的。读者可能对代码中的 Vue.ref()感到疑惑,这是一个响应性函数,后面会介绍。

提示:推荐使用 setup()函数组合式写法。注意,setup()函数里面,不能使用 this。

10.2.4 响应性函数

单击网页链接时,就会打开一个页面,这就是响应性。在 Vue 应用中,当数据被修改时,页面就会自动发生改变,这也是响应性。要为对象创建响应性状态,需要使用 ref、reactive。

1. reactive

```
const admin = Vue.reactive({
        url: 'admin.html'
})
```

代码创建了一个有 url 属性的响应对象。当代表 url 的值改变时,页面会自动更新。reactive 常用在创建具有响应性状态的结构较复杂类型上。

2. ref

如果希望将字符串'你好,Vue!'变成响应式对象,则可使用 ref。ref 会创建一个其值为"你好,Vue!"的字符串对象,并将其传递给 reactive 变成响应性对象。最后,ref 返回对该响应式对象(只包含一个 value 属性)的引用。ref 常用于结构较简单类型。例如上一节的 welcome:

```
const welcome = Vue.ref('你好,Vue!')
```

获取 welcome 的值,在 Vue 内部需使用 welcome.value,注意 HTML 中访问 welcome 的值并不需要这个 value。而 reactive 创建的响应性对象访问时也不需要 value,直接使用即可。

10.2.5 解构

解构,通俗一点的说法就是将一些属性、函数或方法从其定义中"抠"出来以方便使用。例如:

```
const student = Vue.ref({
        sno: '2106080123',
        sname: '杨过'
})
```

如果没有解构,就需要这样访问 sno 属性,以便在浏览器控制台输出:

```
console.log(student.value.sno)
```

解构后,访问形式就简单些:

```
let {sno, sname} = student.value
console.log(sno)
```

现在,10.2.3 节的 setup()函数可改成这样:

```
setup() {
        const {ref, onMounted} = Vue
        const welcome = ref('你好,Vue!')
        onMounted(() => welcome.value += '渐进式框架!')
        return {welcome}
}
```

解构,是一个很方便的做法。可以将一些对象、函数等放在一个专门的集合里面,方便管理。需要使用某个函数的时候,从集合中解构出来即可。

10.3 基础语法

Vue 使用了基于 HTML 的模板语法,模板泛指任何合法的 HTML。Vue 将模板编译并渲染成虚拟 DOM。

10.3.1 模板语法

1. 插值

格式:

```
{{message}}
```

插值是 Vue 中数据绑定最常见的形式了,可以简单将其理解为将 message 数据的内容"插入"到 HTML 指定位置。实际上这是一种数据绑定,绑定了 message 数据属性的值。当 message 的值发生改变时,插值处的内容会自动发生改变。我们在前面的示例中已多次使用过了。

在插值里面,甚至可以使用 JavaScript 表达式(不能是语句)进行某些处理。例如:

```
<span>{{welcome.indexOf('Vue') > 0 ? welcome + '我要努力学习之!' : welcome}}</span>
```

现在，<span>的内容将变成"你好，Vue！渐进式框架！我要努力学习之！"。

2. 指令

Vue 的指令，通常以 v-为前缀，例如 v-html、v-bind、v-if、v-show 等。

1）v-html

通常情况下，插值语句会将里面的内容解读为纯文本。如果前面的示例中，message 的内容是这样的：

```
'你好, <b>Vue!</b>'
```

页面将会原样输出上述内容，而不是期望的加粗的"Vue！"。v-html 指令可以实现 HTML 内容的输出：

```
<span v-html="welcome.indexOf('Vue') > 0 ? welcome + '我要努力学习之!' : welcome"></span>
```

现在，"Vue！"将粗体显示。

2）v-bind

绑定指令 v-bind 日常使用非常频繁。v-bind 用于动态更新 HTML 元素的状态，当绑定的表达式的值发生改变时，会将这种改变应用到页面元素上。例如：

```
<img v-bind:src="myimg">
```

我们可在 setup()函数里面定义 myimg：

```
let myimg = Vue.ref('image/username.png')
return {myimg}
```

页面上将显示 username.png 图片。下面再举一个稍微综合的示例：一个能够输入文字的文本框。

首先，要求边框、文字颜色、字体大小的样式能够组合变化，例如可只设定边框、文字颜色，也可以只设定边框，或者三种都设定。其次，某些场景下改变的是 size 属性，某些场景下改变的则是 maxlength 属性。也就是说具体设置的是哪个属性，并不确定。我们知道，文本框的 size 是设定外观长度的，而 maxlength 则表示允许输入的最大字符长度。这个要求，显然是动态绑定属性的。先来简单定义三个 CSS 样式名称：

```
<style>
    .myColor { color: #00f; }
    .myFont { font-size: 20px; }
    .myBorder { border: 1px dotted #f00; }
</style>
```

然后，页面上放置一个文本框：

```
<div id="app">
    <input v-bind:class="myStyle" v-bind:[attr]="16">你好,Vue!</input>
</div>
```

最后，使用 Vue 进行处理：

```
Vue.createApp({
    setup() {
        const myStyle = {
            myColor: true,
            myFont: true,
            myBorder: false
        }
        const attr = Vue.ref('size')
        return {myStyle, attr}
    }
}).mount('#app')
```

只要改变 myStyle 对象三个属性的不同值，文本框的样式就会发生组合式变化，这里给文本框绑定的是一个对象 myStyle，通过对象的属性，就能够方便地设置不定数量的 CSS 样式了。

v-bind:[attr]中 attr 是一种动态参数，是根据响应数据 attr 的不同值设置不同的属性。读者可以试着将 Vue.ref('size')修改成 Vue.ref('maxlength')，比较修改前后，在文本框中能够输入的字符数是否不同。

3）语法糖

v-前缀对标识 Vue 行为很有帮助，但也稍显烦琐，Vue 为 v-bind 和 v-on 这两个最常用的指令提供了语法糖来简化写法，直接写一个":"号：

```
<input :class="myStyle" :[attr]="16">你好,Vue!</input>
```

4）style 绑定

有时候，对于一些 HTML 元素的简单修饰，常用 style 直接定义，而非采用 CSS 文件方式，例如：

```
<span style="color:#f00">你好</span>
```

Vue 提供了方便的 style 样式绑定处理，我们将前面的 welcome 修改成这样：

```
<span :style="[style1,style2]">{{welcome}}</span>
```

这个<span>叠加绑定了两个样式对象：style1、style2。再看 setup()函数代码：

```
setup() {
    const {ref} = Vue
```

```
        const welcome = ref('你好,Vue!')
        const style1 = ref({
            background: '#090',
            borderRadius: '3px'
        })
        const style2 = ref({
            fontWeight: 'bold',
            color: welcome.value.indexOf('Vue') > 0 ? '#fff' : '#f5c609',
            fontSize: '1.2em',
            textShadow: '3px 3px 3px #605d5d'
        })
        return {welcome, style1, style2}
    }
```

读者可能觉得用 CSS 直接修饰效果也一样,确实如此。不过请注意粗体代码,这里是由 Vue 根据情况进行控制,有些场景就需要这种处理。读者可运行看看最终效果。

5）template

字符串模板,可用于显示 HTML 内容,该模板将会替换已经挂载的 HTML 元素。

示例:用 template 显示两句古诗。

```
<div id="app">
    <span>{{welcome}}</span>
</div>
<script>
    Vue.createApp({
        setup() {
            const welcome = Vue.ref('你好,Vue!')
            return {welcome}
        },
        template:`
            <div>
                <img src="image/username.png" width="32" height="32">
                <span>时人不识凌云木,直待凌云始道高</span>
            </div>`
    }).mount('#app')
</script>
```

页面上不会显示"你好,Vue!",而是显示一张图片和那两句诗。

注意:template 内容开始<div>前面的引号不是英文的单引号"'",也不是中文的"'",而是键盘数字 1 旁边的"`"符号!

10.3.2 计算属性和侦听

1. 计算属性 computed

computed 常用于简单计算。当数据发生改变时,计算属性会自动重新执行并刷新页面数据。

下面通过示例来学习其应用。

示例 1：对 count 进行赋值、取值运算。

```
const count = ref(5)
const comp = computed({
    get: () => count.value++,
    set: val => count.value = val - 20
})
console.log(count.value)      ①
comp.value = 10
console.log(count.value)      ②
return {count}
```

说明：

① 输出 5。数据并没发生改变,所以日志输出 5。

② 输出 -10。给 comp 赋值为 10,将触发 set() 函数进行计算,传递给 val 的值为 10,所以日志输出为 -10。

如果在页面用插值显示 count 的值：

```
<span>{{count}}</span>
```

则显示 9,因为这时会触发 get() 函数,执行 count.value++。

示例 2：计算图书总金额。

```
<div id="app">
    总金额：<span>{{total}}</span>
</div>
<script>
    Vue.createApp({
        setup() {
            const {ref, computed} = Vue
            const books = ref([
                {
                    bname: 'Java Web 开发',
                    price: 48,
                    count: 20
                },
                {
                    bname: 'PostgreSQL 技术及应用',
                    price: 52,
                    count: 18
                }
            ])
            const total = computed(() => {
                let sum = 0
                books.value.forEach(book => {
```

```
                    sum += book.price * book.count
                })
                return sum
            })
            return {total}
        }
    }).mount('#app')
</script>
```

先定义了一个对象数组 books,然后迭代 books 中的每个对象,计算总金额,返回具体值给 total。

2. 侦听 watch

对指定的数据源进行侦听,以便做出反应。基本用法如下:

```
const count = ref(0)
watch(count, newCount => {
    //做出反应,执行处理
    }
)
```

这里对单个数据 count 进行监听,一旦其值发生变化,就可以自动进行处理。

示例:对上一节的示例进行修改。默认是两本书,在文本框输入数字时,会自动显示增减书的数量,并计算总金额,如图 10-2 所示。

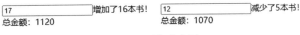

图 10-2　增减书量

当然,这个示例是演示性质,主要是为了学习 watch 的用法,实际中并不会真按照这样的处理逻辑进行操作。

下面是具体代码:

```
<div id="app">
    <input type="number" v-model="bookCount.count" min="1"/>
    <span>{{message}}</span><br>
    总金额:<span>{{total}}</span>
</div>
<script>
    Vue.createApp({
        setup() {
            const {ref, reactive, computed, watch} = Vue
            const books = ref([
                ……
                books.value.forEach(book => {
                    sum += book.price * book.count
                })
```

```
                    return sum
            })
            let message = ref(null)
            const bookCount = reactive({count: books.value.length})   //初值为数组大小
            watch(
                [books, () => bookCount.count],                        ①
                ([books, count], [oldBooks, oldCount]) => {            ②
                    let num = Math.abs(count - oldCount)               //增减数量
                    for (let i = 0; i < num; i++) {
                        count > oldCount ?                             ③
                            books.push({
                                bname: '测试书',
                                price: 10,
                                count: 1
                            }) : books.splice(-1, 1)
                    }
                    if (count > oldCount)
                        message.value = '增加了' + num + '本书!'
                    else
                        message.value = '减少了' + num + '本书!'
                }
            )
            return {bookCount, message, total}
        }
    }).mount('#app')
</script>
```

说明:

① 侦听两个数据:books、count,需要用数组表示。由于bookCount是个响应性对象,所以用箭头函数返回count属性。

② 这是一个箭头函数,传入的是:被侦听的两个数据的新旧状态值。由于两个数据都有新旧两种数据状态,所以需要两个数组。第一个数组,对应这两个数据的当前值,元素顺序需要与被侦听数据的顺序一致。第二个数组,对应这两个数据的旧值。同样,第二个数组中的元素顺序也要与被侦听数据的顺序一致。

③ 用三元运算进行判断处理。若增加书的数量,则向books数组中新增一本书名为"测试书"的对象;否则减少书的数量,从books数组中删除一本书。

当在文本框中输入不同数字时,由于文本框的值与count用v-model双向绑定了,其变化会引起count值的变化,从而触发watch函数,进行处理。关于v-model,下面将会介绍。

注意: 如果输入负数,计算会出现问题。代码并没有进行处理逻辑的严密设计。

10.3.3 表单域的数据绑定

表单域包含了text(文本框)、textarea(多行文本框)、checkbox(复选框)、radio(单选按钮)、select(下拉选择框)等,用于采集用户输入或选择的数据。

Vue 提供了 v-model 指令,能够实现在这些域元素上的双向数据绑定。下面来看一个包含这几种元素双向绑定的综合性示例:

```html
<div id="app">
    姓名:<input v-model="student.name" placeholder="请输入……"/>
    简介:<textarea v-model="student.intro" placeholder="请输入个人简介……"></textarea>
    <p>爱好:<input type="checkbox" v-model="student.fav" value="football"/>足球
        <input type="checkbox" v-model="student.fav" value="dance"/>舞蹈
        <input type="checkbox" v-model="student.fav" value="art"/>艺术
    性别:<input type="radio" v-model="student.sex" value="male"/>男
        <input type="radio" v-model="student.sex" value="female"/>女
        <select v-model="student.major">
            <option value="0701">信息管理</option>
            <option value="0702">电子商务</option>
        </select>
    </p>
    <p>{{student.name}}</p> <!-- 这句及下面这句是为了测试双向绑定效果 -->
    <p style="white-space:pre-wrap">{{student.intro}}</p>
</div>
<script>
    const MyApp = {
        setup() {
            const {reactive, watch} = Vue
            const student = reactive({
                name: '',
                intro: '',
                fav: [],         //复选框是多个值,所以需要用数组,而非字符串
                sex: '',         //没有设置默认选中的值
                major: '0701'    //默认选中"信息管理"
            })
            watch(student, newStudent => {
                newStudent.fav.forEach((s) => console.log(s))  //遍历出选中的全部爱好
                console.log(newStudent.sex)
                console.log(newStudent.major)
            })
            return {student}
        }
    }
    Vue.createApp(MyApp).mount('#app')
</script>
```

从上面代码可以总结出一个规律:一般 checkbox、radio、select 的数据绑定常常与 value 属性结合使用。

10.3.4 组件对象的数据绑定

可以通过 components 关键字来声明组件,并进行数据绑定。那么,如何用这种方式输出"你好,Vue!"? 首先,需定义一个组件化对象:

```
const HelloVue = {
    props: {
        welcome: String
    },
    template:
            `<span>{{ welcome }}</span>`
}
```

前面介绍过 props，用来接收外部传入的数据。这里通过 props 接收字符串数据并传给 welcome 属性。然后，通过 template 模板输出到页面上。

现在，构建 Vue 应用：

```
Vue.createApp({
    components: {
        HelloVue
    },
    setup() {
        const hello = Vue.ref('你好,Vue!')
        return {hello}
    }
}).mount('#app')
```

这里最关键的是用 components 关键字声明了该 Vue 应用要使用的组件 HelloVue。如果是多个组件，组件之间用逗号隔开。最后，在页面使用该组件即可：

```
<div id="app">
    <hello-vue v-model:welcome="hello"></hello-vue>
</div>
```

显然，在 HTML 标准里面，并没有 hello-vue 这样的标签！需要注意的是，命名方式由：小写单词加上"-"连接符组合而成。由代码可知：HelloVue 的 welcome 属性绑定的是 setup() 函数中定义的 hello。

10.3.5 事件绑定和触发

1. 事件绑定 v-on

v-on 用于绑定事件监听器。例如：

```
<button v-on:click="doLogin"></button>
```

可以用语法糖：

```
<button @click="doLogin"></button>
```

v-on 支持使用的修饰符如下。

- ◇ .left：单击鼠标左键时触发。
- ◇ .right：单击鼠标右键时触发。
- ◇ .self：单击当前元素时触发。
- ◇ .once：只触发一次。
- ◇ .stop：阻止事件继续传播。
- ◇ .prevent：阻止元素的默认事件处理。

示例 1：使用 v-on 修饰符。

```html
<a href="http://www.hust.edu.cn" @click.prevent="goUrl">华科</a>
<form name="form1" action="usr/issue" @submit.prevent="check">
    <input type="submit" value="开始发送">
</form>
```

单击链接时，默认会跳转到华中科技大学主页，这里阻止了这个默认行为，而是执行 goUrl() 函数。同样，单击表单的"开始发送"按钮，不会执行默认动作 usr/issue，而是执行 check() 函数。

示例 2：先准备两张灯泡图片 bulbon.gif(点亮)、bulboff.gif(熄灭)。单击按钮时，实现灯泡的点亮/熄灭状态切换。

```html
<div id="app">
    <bulb-image :my-img="curImg"></bulb-image>                     ①
    <p><button @click="changeImg">请单击</button></p>
</div>
<script>
    const BulbImage = {
        props: {
            myImg: String
        },
        template:`<img :src='myImg' width='33' height='60'/>`
    }
    Vue.createApp({
        components: {
            'bulb-image':BulbImage                                 ②
        },
        setup() {
            const curImg = Vue.ref('image/bulbon.gif')
            const changeImg = () =>                                ③
                    curImg.value = curImg.value.match('bulbon') ?
                        'image/bulboff.gif' : 'image/bulbon.gif'
            return {curImg, changeImg}
        }
    }).mount('#app')
</script>
```

说明：

① 绑定按钮单击事件，调用 changeImg() 函数。

② 跟前一节的示例有些差异,这里直接定义了组件名称"bulb-image"。
③ 使用三元条件运算,判断 curImg 是否包含 bulbon,以此切换 curImg 的图片值。

2. 事件触发 $emit

$emit 触发当前对象上的各种事件,基本使用格式:

```
$emit(事件名,参数)
```

例如 $emit('toggle'),其中 toggle 是自定义事件名,然后可与具体的某个函数进行绑定。我们来改写上面的灯泡示例:去掉"请单击"按钮,直接单击灯泡图片时实现灯泡点亮/熄灭状态切换。

```
<div id="app">
    <bulb-image :my-img="curImg" @toggle="changeImg"></bulb-image>
</div>
<script>
    const BulbImage = {
        props: {
            myImg: String
        },
        emits: ['toggle'],          //定义事件名数组
        template: <img :src='myImg' @click="$emit('toggle')" width='33' height='60'/>
    }
    Vue.createApp({
        ...
</script>
```

代码@toggle="changeImg"将 toggle 与 changeImg()函数绑定,@click="$emit('toggle')"则绑定了图片的 click 事件。单击图片时,触发 toggle,自然就会调用 changeImg()函数实现图片的切换。

10.3.6 自定义元素 defineCustomElement

defineCustomElement 用于自定义元素,可实现灵活、高度复用的功能性封装。其基本使用格式大家应该并不会感到陌生:

```
const MyElement = Vue.defineCustomElement({
    props: {},
    setup() {},
    template: `...`,
    styles: [/* CSS 样式 */]
    ...
})
```

可以将其作为单独部分而存在,然后在需要使用的地方注册即可:

```
customElements.define('my-element', MyElement)
```

现在，就可在页面使用了：

```
<my-element></my-element>
```

我们用 defineCustomElement 重新实现上一节的灯泡切换处理：

```
<div id="app">
    <bulb-img></bulb-img>
</div>
<script>
    const BulbImage = Vue.defineCustomElement({
        setup() {
            const curImg = Vue.ref('image/bulbon.gif')
            const changeImg = () => curImg.value = curImg.value.match('bulbon') ?
                                    'image/bulboff.gif' : 'image/bulbon.gif'
            return {curImg, changeImg}
        },
        template: `<img :src='curImg' @click="changeImg" class='bulbimg'/>`,
        styles: [`
                    .bulbimg{
                        width:58px;
                        height:66px;
                        box-shadow: 1px 1px 2px #ccc;
                    }
                `]
    })
    Vue.createApp({
        setup() {
            customElements.define('bulb-img', BulbImage)
        }
    }).mount('#app')
</script>
```

代码中的 styles 只适用于 defineCustomElement。可以直接将 CSS 代码封装在自定义元素里面，非常方便。BulbImage 将灯泡图片切换功能、样式、事件处理等，全部封装在一起，且没那么多绑定处理。相较于 10.3.5 节的代码，可读性、复用性强太多，非常值得读者体会、掌握、熟练使用！

10.3.7 条件和列表渲染

1. 条件渲染

1）v-if、v-else-if、v-else

v-if 指令用于有条件地渲染内容，只有 v-if 的值为真时才会被渲染，类似于 JavaScript 的 if、else if、else。下面对 10.3.2 节的示例 2 计算图书总金额进行判断输出。

```
<div id="app">
    总金额：<span>{{total}}</span>
```

```
    <p v-if="total>=500000">超额</p>
    <p v-else-if="total>=100000">正常</p>
    <p v-else>萎缩</p>
</div>
```

2) v-show

v-show 通过改变元素的 CSS 显示属性来控制 HTML 元素的显示或隐藏，例如：

```
<span v-show="role == 'teacher'">欢迎老师加入!</span>
```

当 role 的值为 teacher 时才会显示"欢迎老师加入!"。

2. 列表渲染

v-for 指令通过对数组中的数据进行循环来渲染列表，常常与 in 或 of 配合使用。请看下面的示例：

```
<div id="app">
    <ul>
        <li v-for="school in schools">                                        ①
            {{school.sname}}({{school.sno}})
            <ul>
                <li style='list-style: none' v-for="(m,index) of school.major">  ②
                    {{index + 1}}.{{m}}
                </li>
            </ul>
        </li>
    </ul>
</div>
<script>
    Vue.createApp({
        setup() {
            const schools = Vue.ref([
                {
                    sno: '0301',
                    sname: '传媒学院',
                    major: ["动画","广告","数字媒体","新闻传播"]
                },
                {
                    sno: '0302',
                    sname: '经济学院',
                    major: ["国际贸易","金融学","财政学"]
                }
            ])
            return {schools}
        }
    }).mount('#app')
</script>
```

说明：

① schools 是一个数组，而 school 代表了数组中的每一个对象。

② 这里的 m 代表专业名称，而 index 则表示该专业对应的索引。索引是从 0 开始编号的，这里使用 index＋1 以便更符合日常排序习惯。

打开页面后，效果如图 10-3 所示。

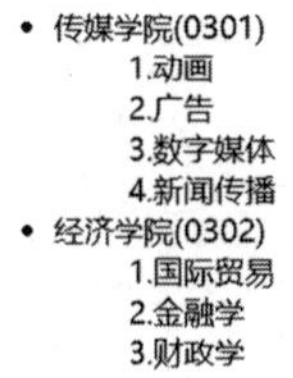

- 传媒学院(0301)
  1. 动画
  2. 广告
  3. 数字媒体
  4. 新闻传播
- 经济学院(0302)
  1. 国际贸易
  2. 金融学
  3. 财政学

图 10-3　渲染后的列表

10.4　h()函数和渲染函数 render()

1. h()函数

h()函数返回一个虚拟节点（VNode），常用来描述需要在页面渲染的 HTML 元素。基本格式如下：

```
h(元素类型,属性,子元素)
```

h()函数可以简单如 Vue.h('span', '你好,Vue!')，或者进行更灵活的组合。

示例：包含图片的层<div>。

```
Vue.h('div', {
    style: {
        width: '50px',
        cursor: 'pointer',
        textAlign: 'center',
        border: '1px solid #00f',
        boxShadow: '2px 2px 3px #ccc'
    }
}, Vue.h('img', {
    src: 'image/bulbon.gif',
    width: '38',
    height: '60',
    title: '点亮前行的路...'
}))
```

代码创建了一个<div>，对其 CSS 样式进行了设置。<div>里面则放置了一张图片，并设置了图片的相应属性。

不过，仅上面的代码还不够，h()函数还需要 render()渲染函数来渲染其返回的 VNode。

2. 渲染函数 render()

渲染函数 render()和 h()函数配合，允许我们充分利用 JavaScript 的编程能力实现自己的目的。值得注意的是，render()函数的优先级高于 template 模板。

示例 1：将上面 h()函数的内容渲染显示。

```
<div id="app"></div>
<script>
```

```
        Vue.createApp({
            render() {
                return Vue.h('div', {
                    ...
                }))
            }
        }).mount('#app')
</script>
```

还可以在setup()函数里面进行渲染：

```
Vue.createApp({
    setup() {
        return () => Vue.h('div', {
            ...
        }))
    }
}).mount('#app')
```

示例2：模拟页面路由。制作会计学院主页内容，当在地址栏打开acc.html时，将显示相关内容。

```
<div id="app">
    <acc-home></acc-home>
</div>
<script>
    const AccHome = Vue.defineCustomElement({
        setup() {
            const home = Vue.h('span',{},'欢迎访问会计学会理事单位会计学院……')
            const routes = {
                '/acc.html': home
            }
            return () => Vue.h(routes[location.pathname])
        }
    })
    Vue.createApp({
        setup() {
            customElements.define('acc-home', AccHome)
        }
    }).mount('#app')
</script>
```

由此可见，h()函数可以生成并控制HTML页面元素。某些场景下，甚至可以代替template，使用render()和h()函数配合来渲染、构建页面。

10.5 使用组件

组件能够实现代码的复用,也方便代码的管理。

10.5.1 组件定义及动态化

1. defineComponent 定义组件

定义组件可使用 defineComponent,其参数比较灵活,一般常使用具有组件选项的对象作为其参数。来看下面的示例:

```
<div id="app"></div>
<script>
    const SchoolsComponent = Vue.defineComponent({
        setup() {
            const schools = [
                {id: 'mg', name: '管理学院'},
                {id: 'md', name: '传媒学院'}
            ]
            return {schools}
        },
        template:
            `<ul>
                <li v-for="school of schools">
                    {{ school.name }}({{ school.id }})
                </li>
            </ul>`
    })
    Vue.createApp(SchoolsComponent).mount('#app')
</script>
```

由此可见,我们前面学过的很多元素,都可应用在 defineComponent 中。

2. 动态组件

还可以采用动态组件这种更灵活的方式来切换组件,基本格式:

```
<component :is="module"></component>
```

利用 is 属性切换不同的组件,实现动态绑定效果。只要改变 module 的值,页面就会显示不同组件的内容。下面用动态组件来改写上面的示例:

```
<div id="app">
    <component :is="curModule"></component>
</div>
<script>
```

```
const SchoolsComponent = Vue.defineComponent({
    ...
})
Vue.createApp({
    setup() {
        const curModule = Vue.ref(SchoolsComponent)
        return {curModule}
    }
}).mount('#app')
</script>
```

代码稍加扩充,就可实现多组件之间的切换变化。第12、13章所实现系统的导航菜单切换,就采取了这种动态组件。

10.5.2 异步组件

异步组件defineAsyncComponent只在需要的时候才会从服务器加载模块。基本使用方法如下:

```
const AsyncComp = Vue.defineAsyncComponent(
    () => new Promise((resolve, reject) => {
        resolve({
            template: '<span>异步组件!</span>'
        })
    })
)
app.component('my-async-component', AsyncComp)
```

defineAsyncComponent返回一个Promise对象并利用resolve回调内容。关于Promise相关知识,请参阅第4章4.8节。下面来看一个较详细的示例:

```
const acDescribe = {
    template:
        <div>
            <span>会计学院是中国会计学会理事单位……</span>
            <span style='color:#0000ff'>稍后请欣赏学院视频……</span>
        </div>
}
const AcComponent = Vue.defineAsyncComponent({
    loader: () => new Promise(resolve => {
        setTimeout(() => {                              //延迟10s后播放视频
            resolve({                                    //回调
                template:
                    `<video src="video/acc.mp4" type="video/mp4"
                        preload="preload" controls autoplay="autoplay"/>
            })
        }, 10000)
    }),
```

```
        delay: 0,                                    //显示 acDescribe 之前延迟的毫秒数
        loadingComponent: acDescribe,                //加载要使用的组件
        errorComponent: Vue.h('span', "视频加载失败……")  //出错提示
})
```

上面的代码利用异步组件,先用 10s 的时间,显示会计学院的文字介绍,然后自动播放学院视频。

10.6 函数集

可以将某个或某类组件调用的函数、方法或常量等抽取出来,放在一个单独的函数集合里面,方便管理。当需要使用时,从该函数集解构出来即可。定义函数集的方法如下:

```
const myFuncs = (function (exports) {
    const buttonStyle = {               //样式常量
        width: '220px',
        height: '22px',
        border: '1px solid #f15555'
    }
    const methodA = () => {   ……  }
    const methodB = () => {   ……  }

    exports.buttonStyle = buttonStyle    //对外暴露
    exports.methodA = methodA
    exports.methodB = methodB
    return exports;
}({}))
```

如果需要调用 buttonStyle、methodB,则解构:

```
setup() {
    const { buttonStyle, methodB} = myFuncs
    ...
}
```

可以分门别类地构建不同的函数集,实现模块化管理。

10.7 使用 Axios 请求后端数据

10.7.1 Axios 简介

Axios 是一个基于 Promise 的网络请求库。Axios 支持 Promise API,能够转换 HTTP 请求并回调响应数据,也能够对请求和响应进行拦截,并能自动转换 JSON 数据,也是 Vue 推荐使用的网络请求库。

可以使用 CDN 方式引用 Axios 库：

```
<script src = "https://unpkg.com/axios/dist/axios.min.js"></script>
```

或者，到官网下载 axios-master.zip，解压后将 dist 文件夹下的 axios.min.js、axios.min.map 复制到项目 js 文件夹下，然后在页面引用：

```
<script src = "js/axios.min.js"></script>
```

10.7.2 请求响应结构和错误处理

在向后端发起请求前，有必要了解 Axios 的请求结构、响应结构。

1. 请求结构

Axios 允许对请求进行丰富配置，一些常用配置如下。
◇ url：'/usr/login' 表示请求地址。
◇ method：'post' 表示请求模式，默认是 get。
◇ headers：{"Authorization", token} 表示自定义请求头。
◇ params：{sname：'张三丰'} 表示与请求 URL 一起发送的数据。
◇ data：user 表示作为请求体被发送的数据，例如 user 对象。
◇ validateStatus：status => status >= 200 && status < 413 表示定义后端返回状态码的正常范围，例如 200~413 的状态码正常处理，不在这范围内的当作出错处理。
◇ cancelToken：new axios.CancelToken(function (cancel) {...}) 表示取消请求并进行处理。

2. 响应结构

Axios 请求后的响应包含如下信息。
◇ data：后端返回的数据。
◇ status：服务器响应的 HTTP 状态码，例如 401(状态文本 Unauthorized)。
◇ statusText：服务器响应的 HTTP 状态文本，例如 Unauthorized。
◇ headers：服务器响应头。
◇ config：请求的配置信息。
◇ request：当前请求。

3. 错误处理

Axios 通过 catch 捕获错误，常见形式如下：

```
myaxios.catch(error =>{
    if (error.response) {
        //请求成功发起、服务器也响应了状态码，
        //但状态代码超出了前面定义的 validateStatus 范围。
    } else if (error.request) {
        //请求已成功发起，但服务器没响应。
    } else {
```

```
            //无法发送请求。
    }
})
```

10.7.3 发起请求

可以发起 Get 或 Post 请求。请看示例：

```
const {status} = await axios({
        method: 'post',
        url: 'usr/regist',
        data: {username: '杨过',password: '123456'}
})
if (status == '205')
        message.value = '该用户已被注册,请更改用户名!'
else if (status == '200')
        message.value = '用户注册成功!'
```

发出用户注册请求后,等待 Axios 返回状态码并解构。根据状态码,进行判断处理。

10.7.4 配置拦截器

Axios 可以添加请求或响应拦截器,二者应用形式相似：

```
axios.interceptors.request.use(config =>{...}, error =>{...})     //添加请求拦截器
axios.interceptors.response.use(response =>{...}, error =>{...})  //添加响应拦截器
```

示例：上传文件时,限制待上传文件为 50MB。

```
const myaxios = axios.create();
myaxios.interceptors.request.use(
    config => {
            if (file.size > (1024 * 1024 * 50)) {
                    cancel('文件大小超过限制!')
                    return config
            }
    }, error => Promise.reject(error)
)
```

10.8 场景应用示例

10.8.1 动态增删书目

第 4 章 4.9 节用 JavaScript 实现了书目的动态增删处理。现在改用 Vue 实现,这样就可以比

较、体会差异,更好地理解 Vue 通过数据的改变来驱动页面变化的特点。下面看具体代码。

1. addbook.html 代码(css 文件略去)

```html
<table id = "books">
    <tr>
        <td colspan = "2" nowrap class = "titleTd">
            <span>新增书目</span></td>
    </tr>
    <tr>
        <td nowrap>书 名</td>
        <td nowrap>
            <div v-for = "(book,index) of books">
                <input type = "text" v-model = "book.name" :ref = "setBookRef" autofocus>   ①
                <input type = "radio" :id = "'r' + index" value = "01"
                                    v-model = "book.checked"/>社会科学
                <input type = "radio" :id = "'r' + index" value = "02"
                                    v-model = "book.checked"/>自然科学
                <img src = "image/delete.png" @click = "delBook(index)" title = "删除"
                                    v-show = "index >= 1">                                   ②
            </div>
        </td>
    </tr>
    <tr><td colspan = 2 class = "buttonTd">
        <img src = "image/add.png" @click = "addBook" title = "新增书目"/> 
        <img src = "image/save.png" title = "保存入库"/>
    </td></tr>
</table>
```

说明:

① 添加新书目后,需要将书名文本框自动设置光标焦点,不然每次都需用户单击书名文本框才能输入,不方便。我们用 ref 来给当前书目文本框注册引用信息。这里使用了动态绑定,将其传递给 setBookRef 函数,在该函数里面保存指向当前书名文本框的引用,以便定位该元素来设置焦点。

② 从第 2 本书开始,才需要显示"删除"按钮,所以用 v-show 进行判断。

2. addbook.js 代码

```js
Vue.createApp({
    setup() {
        const {ref, nextTick} = Vue
        const books = ref([{name: '', checked: '01'}])                          ①
        let bookRefs = []                               //存放书名文本框的引用
        const setBookRef = input => bookRefs[books.value.length] = input        ②
        const addBook = async () => {                   //异步执行
            let newBook = {                             //新增的书目
                name: '',
                checked: books.value[books.value.length - 1].checked            ③
            }
```

```
                    books.value.push(newBook)              //新增书目添加到books集合
                    await nextTick()                                                          ④
                    bookRefs[books.value.length].focus()    //当前新增书目文本框获得焦点
                }
                const delBook = index => {
                    let bookName = books.value[index].name
                    if (confirm('确认删除【' + bookName + '】?')) {
                        books.value.splice(index, 1)         //删除当前书目所在行
                        bookRefs.splice(index, 1)            //同时删除书名文本框的引用
                    }
                }
                return {books, setBookRef, addBook, delBook}
            }
}).mount('#books')
```

说明:

① 页面默认打开时的书目。书名为空,01表示默认选中"社会科学"。

② 将指向当前书名文本框的引用存入数组bookRefs,以便后续通过其设置光标焦点。

③ 新增书目类别,复制上一个书目的类别。如果上一个书目是"自然科学"类,则当前新增书目默认也是该类别。

④ 由于ref绑定的引用在初始渲染的时候并不存在,所以无法立即访问,也就无法直接设置光标焦点。因此,需要利用nextTick推迟到下一个DOM更新周期之后执行focus()。

在这个应用中,只需要改变books数组的数据,页面就会发生变化,这就是数据驱动思想的极佳展示。

10.8.2 学生信息查询

我们在第8章前端用jQuery、后端用Servlet实现了学生信息查询,并在第9章前端用jQuery、后端用Spring Boot又一次实现了该查询。现在,前端改成Vue.js再次来实现。后端代码无须变动,只需要修改前端代码。

视频讲解

1. query.html

```
<div id="query">
    <img src="image/username.png">
    <input type="text" v-model.trim="sname" placeholder="待查询姓名...">
    <img @click="doQuery" src="image/search.png" title="查询"/><br>
    <query-students :url="surl"></query-students>
</div>
<script src="js/query.js"></script>
```

2. query.js

```
const QueryComponent = Vue.defineComponent({         //定义查询组件
    props: {
```

```
            url: String
    },
    setup(props) {
        const {toRefs,reactive,watch} = Vue
        const {url} = toRefs(props)                    //将响应式对象转换为普通对象并解构
        const sdata = reactive({
            loading: false,                            //是否正在进行查询的标志
            error: false,                              //是否出错的标志
            noStudent: false,                          //是否查到数据的标志
            students: []                               //存放查询结果数据的数组
        })
        const queryStudents = url => {
            sdata.students = []                        //清空旧数据
            sdata.noStudent = false
            sdata.loading = true                       //启用查询进行提示
            sdata.error = false
            axios.get(url).then(res => {               //Axios 获取数据
                if (res.data.length > 0)
                    res.data.forEach(s => sdata.students.push(s))   //查到数据置入数组
                else
                    sdata.noStudent = true             //未查到数据
            }).catch(() => sdata.error = true)
              .finally(() => sdata.loading = false)    //关闭查询进行提示
        }
        watch(url, newurl => queryStudents(newurl))    //侦听查询姓名的变化
        return {sdata}
    },
    template:
        `
        <span v-if="sdata.loading">正在获取数据,请稍候…</span>
        <span v-if="sdata.noStudent">查无此数据!</span>
        <span v-if="sdata.error">出错,无法获取数据!</span>
        <table v-if="sdata.students.length>0">
            <tr>
                <td>学号</td><td>姓名</td><td>班级</td>
                <td>电话</td><td>地址</td><td>邮编</td>
            </tr>
            <tr v-for="student of sdata.students">
                <td>{{student.sno}}</td><td>{{student.sname}}</td>
                <td>{{student.sclass}}</td><td>{{student.tel}}</td>
                <td>{{student.address}}</td><td>{{student.postcode}}</td>
            </tr>
        </table>
        `
})
const app = Vue.createApp({
    setup() {
        const sname = Vue.ref('')
```

```
            const surl = Vue.ref('')
            const doQuery = () => {
                surl.value = 'stu/query/' + sname.value;
            }
            return {sname, surl, doQuery}
        }
    })
    app.component('query-students', QueryComponent).mount('#query')
```

10.9 场景任务挑战——下拉选择框联动

1. 我们在第 8 章 8.6.1 节用 jQuery 实现了院系、专业两个下拉选择框的数据联动。现在，请用 Vue.js 重新实现该功能：数据必须使用 8.6.1 节中的 schools 数组。

2. 去掉 query.js 中的 template 部分，修改成用 render() 和 h() 函数实现。

第11章 用图形展示数据

Web 应用数据展示的图形化,往往会给用户带来直接、美观的良好体验。本章主要介绍 Apache ECharts 前端图形显示、JFreeChart 后端图形处理等内容。熟悉 ECharts、JFreeChart 的相关知识,可以为创建美观的 Web 数据展示打下很好的技术基础。

11.1 Web 数据的图形可视化

用文本展示数据,容易进行录入、存储、检索等各种处理,较为灵活,但抽象性较强,难以展现细节,人机交互界面不友好,难以从文字窥探全貌。

用表格展现数据,清晰、规整、规范性、概括性强,能够根据需要进行类别项划分,简便易行,但人机交互界面较差,难以展现数据内在的规律性,也难以从整体上把握数据的趋势与规律。

用图形可视化数据,清晰易见,人机交互界面美观,容易发现数据的规律、趋势,可从多维度去观察数据,也容易从全局把握数据总貌。

因此,Web 数据的图形可视化受到普遍欢迎,得到越来越广泛的应用。

11.2 Apache ECharts 图形前端

11.2.1 Apache ECharts 简介

Apache ECharts(Enterprise Charts)是一个免费、开源、功能强大的图表可视化库,源自百度公司的商业级数据图表工具,ECharts 最早于 2013 年 6 月发布 1.0 版本,得到了业界高度关注和好评,成为国内关注度很高的开源项目,也受到了国外技术团体的关注。

ECharts 用纯 JavaScript 编写,能够实现直观、交互式和高度可定制的图表,现已成为 Apache 基金会孵化项目。

11.2.2　下载与引用

可以通过 CDN 方式，在页面加入以下代码进行引用：

```
<script src = "https://cdn.jsdelivr.net/npm/echarts@5.1.2/dist/echarts.min.js"
        integrity = "sha256-TIOrIaxop+pDlHNVI6kDCFvmpxNYUnVH/SMjknZ/W0Y="
        crossorigin = "anonymous"></script>
```

或者，到官网下载 echarts.min.js，然后复制到项目的 js 文件夹下，再在页面引用：

```
<script src = "js/echarts.min.js"></script>
```

11.2.3　ECharts 创建图形的架构

ECharts 创建图形的方式非常规范，有固定模式可以遵循。

1. 定义一个显示图形的层

```
<div id = "myChart"></div>
```

2. 创建 ECharts 图形对象

```
let chartDom = document.getElementById('chart')     //获取 myChart 层
const myChart = echarts.init(chartDom)              //ECharts 初始化
myChart.setOption(chartsOption)                     //利用 chartsOption 对象细节化
```

3. 对图形进行细节化修饰

chartsOption 是一个 JSON 对象，可从标题、提示文本、图例、背景表格、x 轴、y 轴、series 数据序列等方面进行属性设置、CSS 修饰、数据设定等细节化处理。各项的参数设置很丰富，这里就不一一列出了，感兴趣的读者请参阅 ECharts 说明文档。chartsOption 基本架构如下：

```
chartsOption:
    {
        title: {    //设置图形标题及样式},
        tooltip: {  //鼠标在图形上悬浮时的提示文本},
        legend: {   //图例},
        grid: {     //背景表格},
        xAxis: {    //x 轴},
        yAxis: {    //y 轴},
        series: []  //数据序列
    }
```

并不是每种类型的图形都需要这七大部分，视图形的具体类型而定。

示例：用饼图来显示管理学院 2021 年各专业的招生数据，效果如图 11-1 所示。

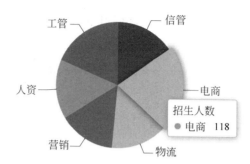

图 11-1　招生饼图

这里并没有从后台数据库中获取数据,而是使用了临时数据。

```
<div id="chartPie" style="margin:0 auto;width:680px; height:400px"></div>
<script>
    const chartPie = Vue.createApp({
        setup() {
            const chartsOption = {
                title: {
                    text: '管理学院 2021 年各专业招生情况一览',
                    left: 'center',
                    top: 5
                },
                tooltip: {
                    trigger: 'item'              //鼠标悬浮在项目上时弹出提示
                },
                legend: {
                    orient: 'horizontal',
                    left: 'center',              //水平放置图例
                    top: 36
                },
                series: [
                    {
                        name: '招生人数',
                        type: 'pie',
                        radius: '50%',           //饼图半径
                        data: [                  //具体数据
                            {value: 83, name: '信管'},
                            {value: 118, name: '电商'},
                            {value: 80, name: '物流'},
                            {value: 81, name: '营销'},
                            {value: 83, name: '人资'},
                            {value: 102, name: '工管'}
                        ],
                        emphasis: {              //鼠标悬浮到饼图分块时强调显示的样式
```

```
                                    itemStyle: {
                                        shadowBlur: 10,
                                        shadowOffsetX: 0,
                                        shadowColor: 'rgba(0, 0, 0, 0.5)'
                                    }
                                }
                            }
                        ]
                    }
                    let chartDom = document.getElementById('chartPie');
                    let myChart = echarts.init(chartDom);
                    myChart.setOption(chartsOption);
                }
            }).mount('#chartPie')
</script>
```

11.3 JFreeChart 图形后端

ECharts 是将后端的数据在前端生成图形，而 JFreeChart 则直接在后端生成图形，然后输出到前端。

11.3.1 JFreeChart 简介

JFreeChart 是一个免费的、纯 Java 的图表库。JFreeChart 提供了丰富的 API，支持 PNG、JPG、SVG、PDF 等多种图表文件格式。

JFreeChart 设计灵活、易于扩展，开发人员可以轻松地在其应用程序中显示高质量的专业图表。

11.3.2 加入 JFreeChart 相关依赖

在 pom.xml 中加入以下代码：

```
<dependency>
    <groupId>org.jfree</groupId>
    <artifactId>jfreechart</artifactId>
    <version>1.5.3</version>
</dependency>
<dependency>
    <groupId>org.jfree</groupId>
    <artifactId>jfreesvg</artifactId>
    <version>3.4.1</version>
</dependency>
```

11.3.3 JFreeChart 应用基础

JFreeChart 提供了两大功能包：chart 图形包、data 数据包。chart 图形包提供了诸如 JFreeChart

（工厂类）、ChatUtils（工具类）、ChartColor（颜色设置）、LegendItem（图例项目）、PiePlot（饼图布局）、CategoryPlot（类别布局）、CategoryAxis（类别坐标轴）、LayeredBarRenderer（直方图渲染器）、TextTitle（文本标题）等常用类。而 data 数据包则提供了 DefaultCategoryDataset（类别数据集）、PieDataset（饼图数据集）、DataUtils（数据工具类）、JSONUtils（JSON 工具类）等常用类。

JFreeChart 的使用比较简单，基本使用模式如下（以饼图为例）：

1. 构建图形所用数据

用 Map 键值对方式存放数据：

```
Map<String, Double> data = new HashMap();
        …
return data;
```

2. 设置图形类别数据集

```
private PieDataset setPieDataset(Map<String, Double> data) {
        DefaultPieDataset dataset = new DefaultPieDataset();
            …
        return dataset;
}
```

3. 建立图形布局并进行修饰

```
PiePlot plot = new PiePlot(dataset);
plot.setSectionOutlinesVisible(true);
…
```

4. 创建 JFreeChart 对象

```
JFreeChart chart = new JFreeChart(title,font,plot, false);
```

5. 输出到具体图形

JFreeChart 提供了 writeBufferedImageAsPNG()、writeBufferedImageAsJPEG()等方法，用于将图形数据输出到图形文件：

```
BufferedImage image = chart.createBufferedImage(690, 420);
ChartUtilities.writeBufferedImageAsPNG(response.getOutputStream(),image, true, 0);
response.flushBuffer();
```

但是，更好的做法是输出到 SVG（Scalable Vector Graphics）图形文件。SVG 是可缩放的、常用于网络的矢量图形，使用 XML 格式定义。SVG 尺寸更小、可压缩性更强，在放大或改变尺寸的情况下图形质量不会下降，且能够被非常多的工具读取和修改。

利用 SVGUtils.writeToSVG()方法，可将数据输出到 SVG 图形文件。

11.4 场景应用示例

为了举例方便,需要创建三个表:school(院系表,包含院系编号、院系名称等字段)、major(专业表,包含专业编号、专业名称等字段)、enroll(招生情况表,包含 id、年份、院系编号、专业招生数据等字段),分别如图 11-2～图 11-4 所示。需要特别说明的是,enroll 表的 major 专业招生数据字段是一个 jsonb 类型,以"专业代码:招生数"数据对的 JSON 格式存放某学院各专业招生情况,如图 11-5 所示。这样设计的好处是:只需要一个字段,就能够将某学院各专业的招生数据存放起来。这些表后续还会用到。

图 11-2　school 表结构

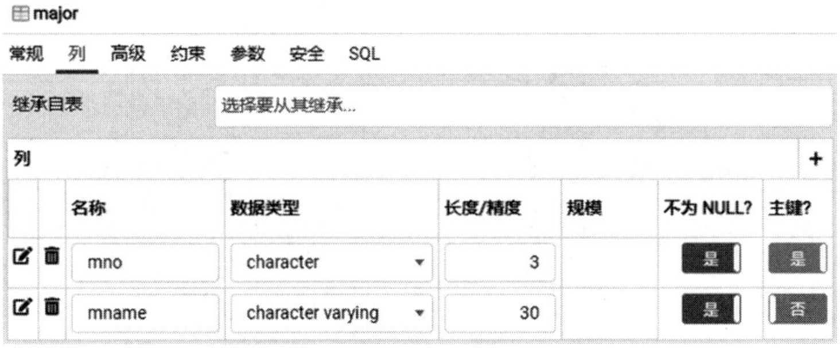

图 11-3　major 表结构

图 11-4　enroll 表结构

数据输出 解释 消息 通知

	id [PK]	nian smallint	cno character(3)	major jsonb
1	1	2021	01	{"011": 88, "012": 82, "013": 85, "014": 84, "015": 101, "016": 83}
2	2	2021	02	{"021": 95, "022": 72, "023": 88}
3	3	2021	03	{"031": 84, "032": 79, "033": 68, "034": 75, "035": 80, "036": 93}
4	4	2021	04	{"041": 79, "042": 83, "043": 82, "044": 75}
5	5	2020	01	{"011": 81, "012": 80, "013": 89, "014": 78, "015": 111, "016": 76}

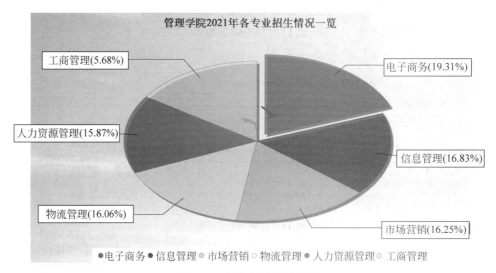

图 11-5 enroll 表的数据

11.4.1 招生情况 SVG 饼图(JFreeChart)

1. 应用需求

用 JFreeChart 输出管理学院(学院编号 01)当前年度各专业招生情况饼图，具体要求：

(1) 从数据表中获取当前年度的数据；

(2) 使用 SVG 图形文件格式，绘制的 SVG 文件存放在 static/image 下面，例如 static/image/mg_enroll_2021.svg；

(3) 饼图背景采用背景图片 static/image/timg.png 进行美化。在浏览器中的效果如图 11-6 所示。

图 11-6 管理学院招生情况

2. 前端：chart_pie.html

```
<div id="chartPie">
        <img :src="enrollImage" v-if="enrollImage!= 'image/'"/>
        <span v-show="enrollImage == 'image/'">无本年招生数据!</span>
</div>
<script>
        Vue.createApp({
```

```
        setup() {
            const enrollImage = Vue.ref(null)
            axios.get('chart/jmg').then(res => enrollImage.value = 'image/' + res.data)
            return {enrollImage}
        }
    }).mount('#chartPie')
</script>
```

<img>绑定的是 enrollImage,其值为后台返回的图片文件名称。如果无招生数据,后台返回空字符。

3. 后端：Enroll 实体类

Enroll 实体类代表了管理学院各专业的招生情况。为了简化问题处理,并未设计成与数据表字段的结构一致。代码如下：

```
@Data
@NoArgsConstructor
@AllArgsConstructor
public class Enroll implements Serializable {
    @Id
    private int id;                    //记录号
    private int nian;                  //年份
    private String school;             //学院
    private String major;              //专业
    private String enrollment;         //招生人数
}
```

4. 后端：EnrollRepository 数据访问接口类

```
public interface EnrollRepository extends R2dbcRepository< Enroll, String > {
    String query = "select c.mname as school,b.value as enrollment " +
            "from enroll a,jsonb_each(a.major) b,major c " +
            "where a.nian = extract(year from now()) and a.cno = '01' and b.key = c.mno";
    @Query(query)
    Flux< Enroll > enrollMG();
}
```

代码的关键是利用 PostgreSQL 数据库的 jsonb_each()函数,遍历出 major 字段数据的键值对(例如: "011":88)集合,然后利用 value 属性(b.value)取出了专业招生数(例如: 88)。b.key 则是指 major 字段数据中每个键值对数据的"键"(例如"011")。

5. 后端：EnrollDao 业务逻辑处理类

```
@Repository
@RequiredArgsConstructor
public class EnrollDao {
```

```java
private final @NonNull EnrollRepository eRepository;            //注入 EnrollRepository
private int curYear = Calendar.getInstance().get(Calendar.YEAR); //获取当前年份

public Mono<String> enrollJMG() {
    Mono<String> m = null;
    try {
        m = buildJPieChart();                   //构建 SVG 饼图,并返回构建成功的文件名
    } catch (Exception e) {
        e.printStackTrace();
    }
    return m;
}
```

现在,向类中加入 buildPieChart()方法:

```java
private Mono<String> buildJPieChart() throws IOException {
    Map<String, Double> data = getData();               //调用 getData()获取数据库中的数据
    if (data.size() == 0)                                //若数据表中无数据,返回空
        return null;

    PieDataset pdata = setPieDataset(getData());         //获取数据集
    PiePlot plot = new PiePlot(pdata);                   //构建饼图布局
    plot.setBackgroundAlpha(0.0f);                       //显示图形的区域背景透明
    plot.setCircular(false);                             //椭圆形
    plot.setLabelGap(0.02);                              //标签间距
    plot.setOutlineVisible(false);                       //不显示边界线条
    plot.setForegroundAlpha(0.7f);                       //图片前景的透明度
    plot.setExplodePercent(pdata.getKey(0), 0.1d);       //第一个专业突出显示
    plot.setLabelGenerator(new StandardPieSectionLabelGenerator(
            ("{0}({2})"), NumberFormat.getNumberInstance(),
            new DecimalFormat("0.00%")));                //显示格式:专业名(数据百分比)
    String title = "管理学院" + curYear + "年各专业招生情况一览";
    JFreeChart chart = new JFreeChart(title, JFreeChart.DEFAULT_TITLE_FONT,
            plot, true);                                 //创建 JFreeChart 对象.true:显示图例
    String imagePath = ResourceUtils.getURL("classpath:static/").getPath(); //背景图片路径
    Image bg = ImageIO.read(new File(imagePath + "/image/timg.png"));  //读取背景图片
    chart.setBackgroundImage(bg);                        //设置背景图片
    return writeToImage(chart, imagePath);               //输出到 SVG 图形文件
}
```

再来看看 setPieDataset()和 getData()方法:

```java
private PieDataset setPieDataset(Map<String, Double> data) {
    DefaultPieDataset dataset = new DefaultPieDataset(); //创建饼图数据集对象
    for (String s : data.keySet()) {                     //遍历 Map 中的键值对数据
        dataset.setValue(s, data.get(s));
    }
```

```java
        dataset.sortByValues(SortOrder.DESCENDING);            //招生数据百分比降序排列
        return dataset;
}
public Map<String, Double> getData() {
    Map<String, Double> data = new HashMap();
    Flux<Enroll> enrolls = eRepository.enrollMG();              //查询招生数据
    List<Enroll> list = enrolls.toStream().collect(Collectors.toList());//转换为 List
    for (Enroll e : list) {                                     //遍历并存入 data 中
        data.put(e.getSchool(), Double.parseDouble(e.getEnrollment()));
    }
    return data;
}
```

最后，就剩下 writeToImage() 方法：

```java
private Mono<String> writeToImage(JFreeChart chart, String imagePath) throws IOException {
    String svgFile = "mg_enroll_" + curYear + ".svg";
    File f = new File(imagePath + "image/" + svgFile);          //创建 SVG 文件对象
    SVGGraphics2D g2d = new SVGGraphics2D(780, 420);            //创建 SVG 图形对象
    Rectangle r = new Rectangle(0, 0, 780, 420);                //绘制区域
    chart.draw(g2d, r);
    SVGUtils.writeToSVG(f, g2d.getSVGElement());                //写入 SVG 文件
    return Mono.just(svgFile);                                  //返回文件名
}
```

6. 后端：EnrollService 服务类

```java
@Service
@RequiredArgsConstructor
public class EnrollService {
    private final @NonNull EnrollDao enrollDao;
    public Mono<String> enrollJMG() {
        return enrollDao.enrollJMG();
    }
}
```

7. 后端：ChartController 控制器类

```java
@RestController
@RequestMapping("chart")
@RequiredArgsConstructor
public class ChartController {
    private final @NonNull EnrollService enrollService;
    @GetMapping("jmg")
    public Mono<String> enrollJMG() {
        return enrollService.enrollJMG();
    }
}
```

11.4.2 招生情况面积图(ECharts)

1. 应用需求

以当前年份为基准,查询管理学院近 5 年各专业招生数据,用面积图显示在页面上。面积图的 x 轴显示专业名称,y 轴为该专业累计招生数,如图 11-7 所示。

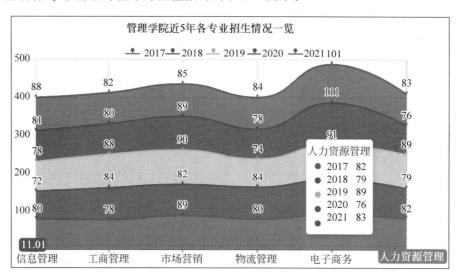

图 11-7 管理学院招生面积图

有一个关键问题需要解决:图形中存在两个数据点:年份、该年份管理学院分专业招生数。我们需要根据图形数据要求的特点,用一种较好的方式从数据库中取出数据。这里采用的取出数据格式如图 11-8 所示。

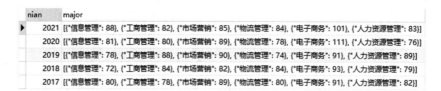

图 11-8 取出的数据集

这种数据格式,很容易与图形要求数据适配。读者可先思考一下如何从数据库中取出这样的数据。下面来看具体实现。

2. 前端:chart_area.html

```
<style>
    #chartArea {
        margin: 0 auto;
        width: 700px;
        height: 400px;
```

```
            padding: 2px;
            border: 1px solid #badaff;
        }
</style>
<div id = "chartArea"></div>
<script src = "js/chart_area.js"></script>
```

3. 前端：chart_area.js

与前面有所不同，这次先定义一个函数集 ChartAreaFunc，里面包含一个常量 chartsOption，用于设置面积图的各种布局参数，还包含一个 buildChartArea() 函数，用来从后台数据库获取数据并构建面积图。然后，在 Vue 应用里面将二者解构出来。整个代码的主体结构如下，比较清晰。

```
const ChartAreaFunc = (function (exports) {
    const chartsOption = {...}
    const buildChartArea = chartsOption => {...}
    exports.chartsOption = chartsOption
    exports.buildChartArea = buildChartArea
    return exports;
}({}))
Vue.createApp({
    setup() {
        const {chartsOption, buildChartArea} = ChartAreaFunc    //解构
        buildChartArea(chartsOption)                            //构建图形
    }
}).mount('#chartArea')
```

1) chartsOption 常量对象

```
const chartsOption = {
    title: {
        text: '管理学院近5年各专业招生情况一览',
        left: 'center',
        top: 5,
        textStyle: { color: '#0a38ef' }
    },
    tooltip: {
        trigger: 'axis',
        axisPointer: {
            type: 'cross',           //鼠标滑动时显示交叉线
            label: { backgroundColor: '#ea0a23' }
        }
    },
    legend: {
        top: 36,
        data: [                      //近5年
            (new Date().getFullYear() - 4) + '',
```

```
                        (new Date().getFullYear() - 3) + '',
                        (new Date().getFullYear() - 2) + '',
                        (new Date().getFullYear() - 1) + '',
                        new Date().getFullYear() + '']
            },
            grid: {
                    left: '0px',
                    right: '40px',
                    bottom: '2px',
                    containLabel: true
            },
            xAxis: [ {
                    type: 'category',
                    boundaryGap: false,             //不扩展坐标轴两端空白
                    splitLine: {
                            show: true              //显示分割线
                    },
                    data: []                        //x 轴的数据,在 buildChartArea()函数中赋值
            } ],
            yAxis: [ {
                    type: 'value',
                    splitLine: {
                            show: true
                    }
            } ],
            series: []                              //数据序列,在 buildChartArea()函数中赋值
}
```

2) buildChartArea()函数

```
const buildChartArea = chartsOption => {
    //5 个年份面积图的对应颜色
    const areaStyleColor = ['#0886f5', '#cb0dec', '#f3c007', '#ea0a23', '#099109'];
    axios.get('chart/emg').then(res => {                    //后台获取数据,请求路径:chart/emg
        if (res.data.length > 0) {
            res.data.forEach((o, index) => {
                let seriesObject = {                        //序列数据对象
                    name: o.nian,
                    type: 'line',
                    stack: '招生',
                    smooth: true,
                    symbol: 'pin',                          //折线上显示标注图标
                    symbolSize: 6,                          //圆点大小
                    label: {show: true},                    //圆点上显示数字
                    color: areaStyleColor[index],           //圆点颜色
                    lineStyle: {                            //线条颜色
                            color: areaStyleColor[index]
```

```javascript
                    },
                    areaStyle: {                                    //面积区颜色
                        color: areaStyleColor[index]
                    },
                    emphasis: { focus: 'series' },
                    data: []
                }
                let majors = JSON.parse(o.major);                   //专业数据解析成JSON对象
                majors.forEach(major => {
                    for (let key in major) {                        //处理每个专业
                        seriesObject.data.push(major[key]);         //招生数据存入数组
                        //若专业名称未曾加入x轴,则加入
                        if (chartsOption.xAxis[0].data.indexOf(key) == -1) {
                            chartsOption.xAxis[0].data.push(key)
                        }
                    }
                })
                chartsOption.series.push(seriesObject)
            });
            let chartDom = document.getElementById('chartArea')
            let myChart = echarts.init(chartDom)
            myChart.setOption(chartsOption)
        }
    }).catch(err => console.log(err))
}
```

4. 后端：EnrollRepository 数据访问接口类

加入以下代码,查询出管理学院近5年各专业招生数据：

```java
String qry = "select a.nian,jsonb_agg(jsonb_build_object(c.mname,b.value))::text as major " +
        "from enroll a,jsonb_each(a.major) b,major c where a.cno = '01' " +
        "and a.nian >= extract(year from now()) - 4  " +
        "and b.key = c.mno group by a.nian order by a.nian";
@Query(qry)
Flux<Enroll> enrollMG5();
```

5. 后端：EnrollDao 业务逻辑处理类

加入以下代码：

```java
public Flux<Enroll> enrollMG5() {
        return eRepository.enrollMG5();
}
```

6. 后端：EnrollService 服务类

加入 enrollMG5()方法：

```
public Flux < Enroll > enrollMG5() {
        return enrollDao.enrollMG5();
}
```

7. 后端：ChartController 控制器类

加入 emg 映射：

```
@GetMapping("emg")
public Flux < Enroll > enrollMG5() {
        return enrollService.enrollMG5();
}
```

11.5 场景任务挑战——招生情况直方图

以 11.4.2 节为基础，不修改后端程序代码，新建页面 chart_bar.html，用 ECharts 显示管理学院近五年招生情况直方图，如图 11-9 所示。

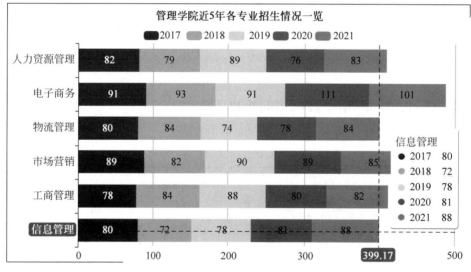

图 11-9 管理学院招生直方图

第 12 章

消息服务

消息服务(Message Service,MS)在现实中得到了广泛应用。本章主要介绍了消息服务模式、传递方法、Apache Kafka、WebSocket、SockJS 以及 Spring WebFlux 对 WebSocket 的处理思路等内容,并用一个聊天室综合了这些知识的具体使用。掌握消息服务的内容,有利于实现 Web 消息的推送和数据交换处理。

12.1 消息服务概述

12.1.1 消息服务简介

消息(Message)是指在应用之间传送的数据,可以是普通的文本字符串,也可以是一个对象。消息服务是指在应用之间提供消息传递并进行消息管理。

用户发送或接收消息,消息服务器(或称消息件、消息处理平台)接收用户发送的消息,存储消息并转发给各类应用(例如日志、短信提醒、账务提示、订单提醒等),也可视需要存入数据库。当然,消息服务器也可以接收来自各类应用发送的消息,再转发给用户。消息服务结构如图 12-1 所示。

图 12-1 消息服务结构

一般消息服务器作为中间桥梁,需要具备异步、平台独立、配置方便等特点。可靠性是衡量消息服务器的非常重要的指标,不少消息处理平台宣称"零消息丢失"。

12.1.2 消息服务模式

一般将发送消息方称为生产者,接收消息方称为消费者。消息服务有两种常见模式:点对点模式、发布订阅模式。

1. 点对点模式(Point to Point,P2P)

消息通过虚拟的消息队列(Message Queue,MQ)进行交换。MQ 是一种应用之间消息通信的方式。消息队列中的某条消息只能被一个接收者消费。生产者和消费者之间是松散的,可在运行时动态添加,就像甲随时可以向乙发送消息,并不需要甲、乙事先约定好。甲发送消息后可以等待乙的回应,也可以无须乙的回应。

消费者从消息队列中获取到消息后,该消息就会被从队列中移除。消费者不能再去队列中获取该消息了,也就是说消费者不能消费已被消费过的消息。

2. 发布订阅模式(Publish/Subscribe)

生产者将消息发送到主题(topic)中。希望获取某个主题消息的消费者,需要订阅该主题才能接收到生产者发布的消息。一个主题上可以有许多订阅者,每个订阅者都能收到生产者发布的消息,即发布到 topic 的消息可被所有订阅者消费,每个订阅者都能得到一份消息的备份。发布订阅模式的结构如图 12-2 所示。

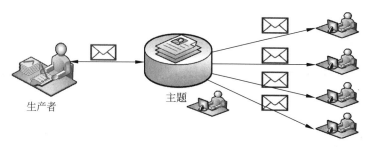

图 12-2　发布订阅模式

3. 消息传递方法

以 JSON 格式发送或接收消息。主要采用两种方法:

◇ 通过轮询收发消息。所谓轮询就是客户端定时向服务器发出请求,不管服务器有无结果返回,到下一个时间点继续下一轮的轮询。这种方式容易造成带宽和服务器资源的浪费。当客户端数量较多时,容易出现问题。

◇ 通过 WebSocket 收发消息。WebSocket 是一种标准化方法,客户端和服务器只需要一次 HTTP 握手,通过单个 TCP 连接,通信过程建立在全双工双向通信通道中。连接建立后,服务器端可主动推送消息到客户端,直到客户端关闭请求。客户端无须循环发出请求。WebSocket 能够实现消息的实时通知,性能开销小,节省服务器资源和带宽。非常适合 Web 应用程序中客户端和服务器需要高频率、低延迟、大数据量交换的场景,例如实时数据采集、聊天、金融股票展示等方面。

4. 流行的消息队列产品

市场上的消息队列产品很多,例如 RabbitMQ、ActiveMQ Artemis、Kafka 等。

RabbitMQ 是一个用 Erlang 语言开发的、开源的消息队列服务软件,以其高性能、健壮、可伸缩性在业界闻名。

ActiveMQ Artemis 是一款开源、多协议、基于 Java 的高性能、非阻塞的消息服务产品,支持集群、共享存储、JDBC 等特性。

Kafka 则来自著名的 Apache 软件基金会。本章将使用该产品作为消息服务器。

Spring Boot 对上述三种产品都提供了支持。

12.2 用 Apache Kafka 作为消息服务器

12.2.1 Apache Kafka 简介

Apache Kafka 是 Apache 基金会最活跃的五大项目（Flink、Lucene-Solr、Ignite、Kafka、Tomcat）之一，是一种开源、分布式流处理平台。

Kafka 可扩展、高可用、数据持久化、高吞吐量、高性能，宣称"零消息丢失"。Kafka 在国内外市场都得到了广泛应用，互联网 IT 公司、各类厂商、证券交易所，以及成千上万的机构都在用 Kafka 进行数据集成、消息服务应用、高性能数据管道等，在金融、保险、制造、通信等行业有较高的市场占有率。

12.2.2 启用 Kafka 服务器

到官网下载 Kafka，解压到某个文件夹，例如：D:\kafka_2.13-2.8.0。

（1）启动 zookeeper。

进入 DOS 命令提示符方式，改变目录到 D:\kafka_2.13-2.8.0\bin\windows，输入命令：

```
zookeeper-server-start   ../../config/zookeeper.properties
```

启动 zookeeper，在这过程中，Kafka 会创建一个文件夹 d:\tmp，作为运行过程中的临时文件夹。

（2）启动 Kafka。

打开另外一个 DOS 命令提示符窗口，改变目录到 D:\kafka_2.13-2.8.0\bin\windows，输入命令：

```
kafka-server-start   ../../config/server.properties
```

启动 Kafka，如图 12-3 所示。

（3）创建消息主题。

本章的场景应用采用发布订阅模式来收发消息，所以需要创建一个消息主题 schat。再次打开一个 DOS 命令提示符窗口，仍然改变目录到 D:\kafka_2.13-2.8.0\bin\windows，输入命令：

```
kafka-topics  --create  --bootstrap-server localhost:9092  --topic schat
```

Kafka 提示创建完成，如图 12-4 所示。

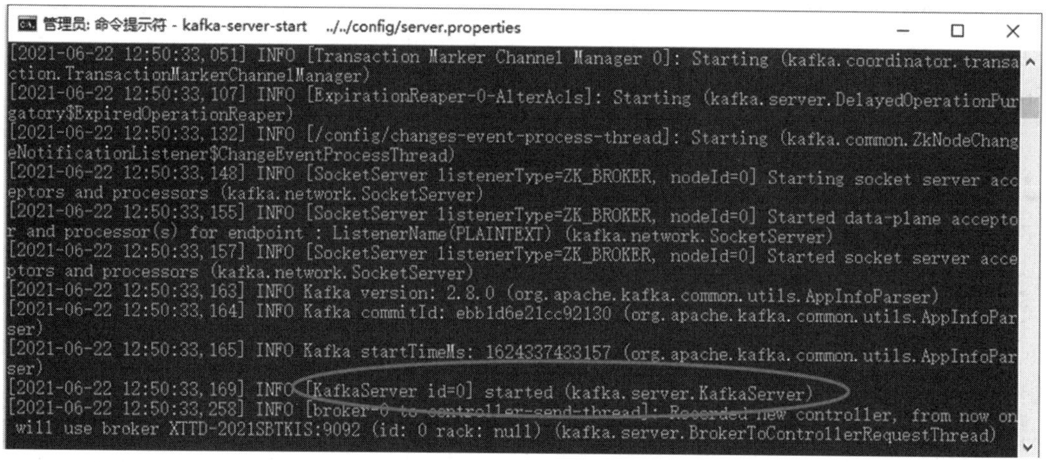

图 12-3 启动 Kafka

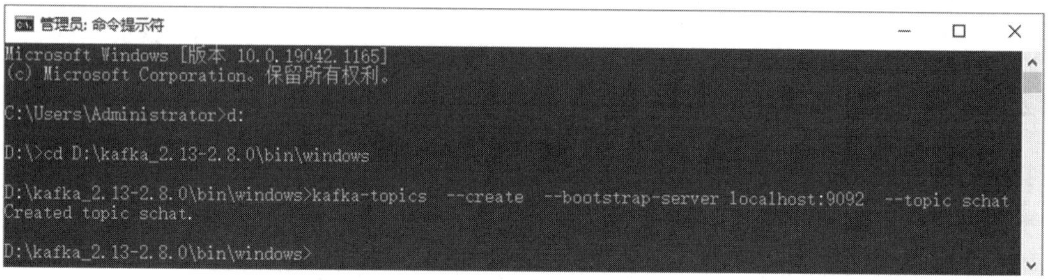

图 12-4 创建主题

12.2.3 Kafka 配置和管理

1. 加入 Kafka 依赖

```
<dependency>
    <groupId>org.springframework.kafka</groupId>
    <artifactId>spring-kafka</artifactId>
</dependency>
<dependency>
    <groupId>io.projectreactor.kafka</groupId>
    <artifactId>reactor-kafka</artifactId>
    <version>1.3.4</version>
</dependency>
```

2. 辅助管理软件 Offset Explorer

由于 Kafka 并没有提供图形化的管理工具，实际中使用会有诸多不便。可以借助第三方工具，例如 Offset Explorer、Kafka-Manager、Kafka Monitor 等。Offset Explorer 是一款 Windows 下的实用工具，可以到其官网免费下载。图 12-5 为安装后的消息管理界面。

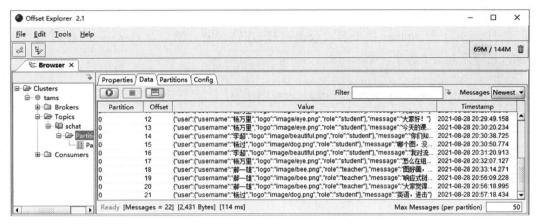

图 12-5　Offset Explorer 管理界面

12.2.4　KafkaTemplate 模板

Spring 提供了 KafkaTemplate，无须进行烦琐的配置，高度封装化。只需要注入 KafkaTemplate 即可：

```
private final @NonNull KafkaTemplate kafka;
```

然后，就可用该模板发送消息：

```
kafka.send(topicName, message.getPayload());
```

接着，就可利用@KafkaListener 注解某个方法，接收消息并处理：

```
@KafkaListener(topics = "schat")
public void processMessage(String message) {
        //对接收到的消息内容 message 进行处理
}
```

12.2.5　生产者 Producer 和消费者 Consumer

1. 生产者 Producer

也可以创建生产者 Producer，利用生产者发送消息。首先，需要注入生产者工厂 ProducerFactory：

```
private final @NonNull ProducerFactory producerFactory;
```

然后，就可以利用该工厂创建生产者并发送消息：

```
Producer producer = producerFactory.createProducer();
producer.send(new ProducerRecord(topicName, "hello,kafka!"));
```

2. 消费者 Consumer

除了利用@KafkaListener 监听器接收消息外,还可以通过消费者 Consumer 接收消息。与生产者类似,先要注入消费者工厂 ConsumerFactory:

```
private final @NonNull ConsumerFactory consumerFactory;
```

再利用该工厂创建消费者,并消费消息:

```
Consumer consumer = consumerFactory.createConsumer();
ConsumerRecords<String, String> records = consumer.poll(Duration.ofMillis(1000));
```

代码在1000ms 内拉取消息并返回消费记录集合,后续就可以遍历该集合,对集合中的消息进行处理。

12.2.6 Kafka 响应式发送器和接收器

1. 发送器 KafkaSender

Kafka 提供了响应式消息发送器 KafkaSender,用来发送消息。KafkaSender 一般需要通过 SenderOptions 配置项来创建,需要先注入配置对象 KafkaProperties:

```
private final @NonNull KafkaProperties kafkaProp;
```

然后,创建 SenderOptions:

```
SenderOptions options = SenderOptions.create(kafkaProp.buildProducerProperties());
```

最后,就可创建发送器并发送消息:

```
KafkaSender.create(options).send(……);
```

2. 接收器 KafkaReceiver

与 KafkaSender 类似,Kafka 提供了响应式消息接收器 KafkaReceiver,用来接收消息。同样,KafkaReceiver 一般需要利用 ReceiverOptions 配置项来创建,例如:

```
private final @NonNull KafkaProperties kafkaProperties;
ReceiverOptions<Integer, String> options =
        ReceiverOptions.<Integer, String>create(kafkaProperties.buildConsumerProperties());
Flux<ReceiverRecord<Integer, String>> flux = KafkaReceiver.create(options).receive();
flux.subscribe(r -> {
    //对消息进行处理
});
```

先注入 KafkaProperties，再利用其构建 ConsumerProperties，基于此创建 ReceiverOptions 对象。然后创建 KafkaReceiver 对象，接收消息。最后，对消息进行处理。

12.3 整合 WebSocket 及 SockJS

12.3.1 在客户端使用

1. SockJS 简介

SockJS 是一个 JavaScript 库。当浏览器不支持 WebSocket 时，SockJS 无须改动代码就可进行模拟支持。也就是说，SockJS 可根据浏览器环境自动选择是采用 WebSocket 还是模拟支持，因而更具有通用性。SpringBoot 亦对 SockJS 提供了良好支持。

要在客户端使用 SockJS，可去 https://github.com/sockjs/sockjs-client 下载 sockjs-client-main.zip，解压后将 sockjs.min.js、sockjs.min.js.map 复制到项目 js 文件夹下，然后在页面引入即可：

```
<script src = "js/sockjs.min.js"></script>
```

2. SocketJS 使用方法

WebSocket 访问格式为 ws://主机/WebSocket 端点，例如 ws://localhost:8080/schat。SockJS 为方便处理进行了包装，其访问地址格式如下。

协议://主机名:端口号/应用名称/WebSocket 端点/{集群中的路由服务器}/{自动生成的会话 ID}/{传输类型}

例如：ws://localhost:8080/schat/token/409/f44q1cht/websocket

这是 SockJS 自动选择、匹配后的地址，其中 409、f44q1cht 由 SockJS 随机生成。我们其实只需要这样写：/schat/token。SockJS 语法格式如下。

```
const sockjs = new SockJS(url, _reserved, options);
```

◇ url：使用常规地址格式，无须使用 ws 协议，例如：/schat/token。
◇ _reserved：内部保留。
◇ Options：参数选项，可指定 server 数据参数、sessionId、timeout 超时毫秒数等，例如 {timeout:20000}。

3. 连接 WebSocket 处理消息

SockJS 提供了连接成功、消息处理、出错处理、连接关闭的常规应用模式：

```
const sockjs = new SockJS('/schat/token', null, {timeout: 20000})
sockjs.onopen = () => {
    //连接成功的处理
```

```
}
sockjs.onmessage = event => {
    //对消息进行处理
}
sockjs.onclose = () => {
    //连接关闭处理
}
sockjs.onerror = error => {
    //出错处理
}
```

要发送消息,则可使用send()函数:

```
sockjs.send(message)
```

提示:如果不使用SockJS,则连接WebSocket的对应方式如下:

```
const sockjs = new WebSocket('ws://localhost:8080/schat/token')
```

12.3.2 在服务端使用

1. 创建 WebSocket 服务器

创建一个类,定义为TextWebSocketHandler的子类,然后改写其三个方法。

```
@Service
public class ChatService extends TextWebSocketHandler {
    // 这里可定义或注入一些变量、对象等......
    @Override
    public void afterConnectionEstablished(WebSocketSession session) {
        // 连接建立后,可在此做些准备工作......
    }
    @Override
    public void handleTextMessage(WebSocketSession session, TextMessage message) {
        // 可在此处理消息......
    }
    @Override
    public void afterConnectionClosed(WebSocketSession session, CloseStatus status) {
        // 可在这里做些清理工作......
    }
}
```

2. 配置 WebSocket 服务器及访问端点

定义一个配置类,启用WebSocket,实现WebSocketConfigurer接口。然后实现registerWebSocketHandlers()方法,在该方法中定义WebSocket端点:

```
@Configuration
@EnableWebSocket
public class ChatConfig implements WebSocketConfigurer {
    //定义或注入一些实例变量……
    @Override
    public void registerWebSocketHandlers(WebSocketHandlerRegistry reg) {
        //注册 WebSocket 服务器、定义端点……
    }
}
```

12.3.3　使用拦截器

拦截器可用于在连接 WebSocket 前进行判断，例如只有有效的登录用户才能发送聊天消息。要使用拦截器，需要自定义一个返回类型为 HandshakeInterceptor 的方法，然后返回一个 HttpSessionHandshakeInterceptor 对象，并改写该对象的 beforeHandshake() 方法实现拦截规则。

```
private HandshakeInterceptor chatInterceptor() {
    return new HttpSessionHandshakeInterceptor() {
        @Override
        public boolean beforeHandshake(ServerHttpRequest request, ServerHttpResponse response,
                WebSocketHandler wsHandler, Map<String, Object> attributes) {
            //可在此进行拦截处理……
        }
    };
}
```

12.3.4　Spring WebFlux 中的 WebSocket

Spring WebFlux 提供了对 WebSocket 的支持，基本处理思路是：

1. 实现 WebSocketHandler 接口

```
public class ChatHandler implements WebSocketHandler {
    @Override
    public Mono<Void> handle(WebSocketSession session) {
        // 进行消息处理……
    }
}
```

这里创建了一个基于 WebFlux 的聊天服务类 ChatHandler。

2. receive() 和 send(Publisher<WebSocketMessage>)

利用 WebSocketSession 的 receive() 方法，就可以接收消息流。而 send() 方法则可发送消息，并返回一个 Mono<Void>。例如：

```
Flux<WebSocketMessage> sFlux = session.receive()          //接收消息
            .map(WebSocketMessage::getPayloadAsText)      //消息体
            .map(session::textMessage);                   //转换为文本
return session.send(stringFlux);                          //发送消息
```

3. 注册映射路径

```
@Bean
public HandlerMapping handlerMapping() {
    Map<String, ChatServiceHandler> map = new HashMap<>();
    map.put("/schat", new ChatHandler());
    return new SimpleUrlHandlerMapping(map, 1);
}
```

利用 SimpleUrlHandlerMapping 即可实现 ChatServiceHandler 服务端点地址"/schat"的注册，方法参数中的 1 表示加载的优先级，像前面学过的 ServerResponse、ResponseEntity 等默认优先级为 0。数值越小优先级越高。

视频讲解

12.4 场景应用示例——聊天室

12.4.1 应用需求

这是一个简易聊天室，如图 12-6 所示。

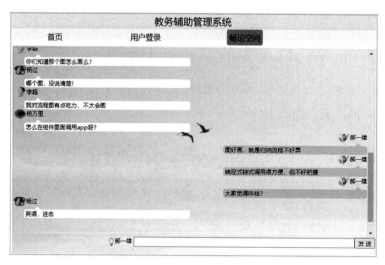

图 12-6 聊天界面

功能要求：

(1) 登录后的用户才能聊天；

(2) 当前用户的聊天内容在右边显示，聊天内容的背景为绿色。其他用户的聊天内容在左侧显示,聊天内容的背景为白色。整个界面用一张图片作为背景。

技术要求：

(1) 前端 Vue.js 实现,并将登录、聊天定义为两个组件,实现模块化处理,后端 SpringBoot 实现；

(2) 使用 SockJS 通过 WebSocket 进行消息传递；

(3) 使用 Kafka 存储消息,并用响应式发送器、接收器收发消息；

(4) 用拦截器验证用户登录状态；

(5) 用户进入聊天室后,自动向用户推送最近的 15 条消息记录。整个技术架构如图 12-7 所示。

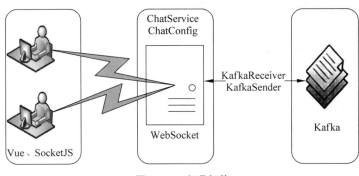

图 12-7　实现架构

12.4.2　主页

1. index.html

```
<div id = "app">
    <div class = "menu - title">教务辅助管理系统</div>
    <div class = "menu - link">
        <ul class = "menu - ul">
            <li v - for = "(item,index) in menus" @click = "menuIndex = index">        ①
                {{item.name}}
            </li>
        </ul>
    </div>
    <div class = "menu - detail">
        <component :is = "curComponent"></component>                                    ②
    </div>
</div>
<script type = "module" src = "index.js"></script>
```

说明：

① menus 存放了代表"主页""用户登录""畅论空间"这三个导航菜单的组件,用一个无序列表显示,并用 CSS 样式进行修饰(代码略去)。单击菜单项时,改变当前菜单的索引 menuIndex,利用 Vue 的侦听器做出反应。

② 动态挂载组件内容,主要是利用了 <component> 的 is 属性。

2. index.js

```
import {UsersLoginComponent as login} from "./js/users.component.js"    //导入组件
import {ChatComponent as chat} from "./js/chat.component.js"

const menus = [                                                         //定义导航菜单使用的组件
    {id: 'index', name: '首页', module: null},
    {id: 'login', name: '用户登录', module: login},
    {id: 'chat', name: '畅论空间', module: chat}
]
const index = {
    setup() {
        const {ref, watch} = Vue
        const menuIndex = ref(null)
        const curComponent = ref(null)
        //侦听菜单索引值的变化,以便加载对应的组件
        watch(menuIndex, (newIndex) => curComponent.value = menus[newIndex].id)
        return {menus, menuIndex, curComponent}
    }
}
const app = Vue.createApp(index)
menus.forEach((item) => {                                               //遍历数组,注册组件
    app.component(item.id, item.module)
})
app.mount('#app');
```

12.4.3 登录组件 users.component.js

```
export const UsersLoginComponent = Vue.defineComponent({
    setup() {
        const {ref, reactive} = Vue
        const state = ref(null)                                                     ①
        const isDoing = ref(false)                              //登录进行状态的标识量
        const user = reactive({username: '', password: ''})     //登录用户
        const doLogin = async () => {
            isDoing.value = true
            state.value = null
            sessionStorage.removeItem("loginer")                //清除临时会话数据
            sessionStorage.removeItem("token")
            user.password = CryptoJS.MD5(user.password) + ''    //对密码进行加密
            try {
                const {data, headers} = await axios({           //解构 axios 返回的数据
                    method: 'post',
                    url: 'usr/login',
                    data: user
                })
```

```
                    if (data == null)
                        return
                    let auth = headers['authorization']
                    if (auth != null) {
                        state.value = 1
                        sessionStorage.setItem('loginer', JSON.stringify(data))      ②
                        sessionStorage.setItem('token', auth)
                    } else
                        state.value = 0
                } catch (error) {
                    console.error(error)
                }
                isDoing.value = false
            }
            return {user, state, isDoing, doLogin}
        },
        template:        //构建登录界面元素
            <div class="user-login">
                账户登录<br>
                <img src="image/username.png">
                <input type="text" class="login-input" placeholder="用户名"
                                            v-model="user.username"><br>
                <br><img src="image/password.png">
                <input type="password" class="login-input" placeholder="密    码"
                                            v-model="user.password">
                <br><br><button class="login-button" @click="doLogin">
                                            登   录</button><br>
                <div v-show="isDoing">
                    <img src="image/progress.gif" width="50" height="50"/>
                                            正在登录,请稍候……
                </div>
                <span style="color:#ff0000" v-show="state==1">用户登录成功!</span>
                <span style="color:#ff0000" v-show="state==0">
                                            用户名或密码错误,登录失败!</span>
            </div>
})
```

说明:

① state 标识登录状态:0,登录失败;1,登录成功。

② sessionStorage 用于临时存储会话信息,以便在不同页面之间共享数据。sessionStorage.setItem()将信息存入指定名称的会话中,sessionStorage.removeItem()则予以清除。用户登录成功后,后台返回两个数据:用户数据(包含用户名、用户头像、用户角色等,也就是 data 中的数据);JWT 令牌(附加在 HTTP 响应头 Authorization 中,通过 headers['authorization']获取得到)。

12.4.4 登录后端处理

大部分与第 9 章相同,只有 UsersController 控制器类略作修改:为简化起见,去掉了 Hazelcast

缓存处理。

```
@PostMapping("login")
public Mono < ResponseEntity > doLogin(@RequestBody Users user) {
    return usersService.login(user)
            .flatMap(u -> {
                …
            });
}
```

12.4.5 聊天组件 chat.component.js

```
export const ChatComponent = Vue.defineComponent({
    setup() {
        const {ref, reactive, onMounted} = Vue
        const chatContent = ref([])                          //聊天内容
        let me = reactive({                                                              ①
            user: {username: '', logo: '', role: ''},
            message: '您还未登录,请登录后畅论……\n'
        })
        const default_msg = {
            divclass: 'welcome_msg',
            msgclass: '',
            content: {
                user: {username: '', logo: ''},
                message: '连接成功,开始畅论……\n'
            }
        }
        const token = sessionStorage.getItem("token")      //获取登录时存储的令牌
        onMounted(() => {
            if (token != null) {
                const usr = JSON.parse(sessionStorage.getItem("loginer"))
                me.user.username = decodeURIComponent(usr.username)                      ②
                me.user.logo = usr.logo
                me.user.role = usr.role
                me.message = ''
                bindListener()
            }
        })
        let sockjs = null
        const bindListener = () => {
            sockjs = new SockJS('/schat/' + me.user.username + '/' + token, null,
                                    {timeout: 20000})
            sockjs.onopen = () => {                          //连接成功显示欢迎信息
                chatContent.value.push({
                    divclass: 'welcome_msg',
```

```js
                            msgclass: '',
                            content: {
                                user: {username: '', logo: ''},
                                message: '连接成功,开始畅论……\n'
                            }
                        })
                    }
                    sockjs.onmessage = event => {                  //处理聊天消息
                        let chatData = {divclass: 'div_left', msgclass: 'msg_left', content: {}}   ③
                        let msg = JSON.parse(event.data)           //获取后台推送的聊天消息
                        if (msg.user.username === me.user.username) {  //当前用户右侧显示样式
                            chatData.divclass = 'div_right'
                            chatData.msgclass = 'msg_right'
                        }
                        chatData.content = msg                     //后台返回的聊天消息
                        chatContent.value.push(chatData)           //存入 chatContent
                        let chatArea = document.getElementById('chat-area')
                        chatArea.scrollTop = chatArea.scrollHeight //自动滚动到聊天窗底部
                    }
                    sockjs.onclose = () => {
                        default_msg.content.message = '与消息服务器的连接已关闭!\n'
                        chatContent.value.push(default_msg)
                    }
                    sockjs.onerror = () => {
                        default_msg.content.message = '出错,无法连接消息服务器!\n'
                        chatContent.value.push(default_msg)
                    }
                }
                const sendMsg = () => {                            //发送消息
                    if (token == null)
                        return
                    sockjs.send(JSON.stringify(me))                //JSON 字符串格式发送
                    me.message = ''                                //发送后,聊天文本框置空
                }
                return {chatContent, me, sendMsg}
            },
            template: `
                <div class="chat">
                    <div id="chat-area">
                        <ul class="chat_ul">
                            <li v-for="(chat, index) in chatContent">
                                <div :class='chat.divclass'>
                                    <img v-if="chat.content.user.logo!= ''" :src='chat.content.user.logo'
                                        width=26 height=20 >{{chat.content.user.username}}
                                    <div :class='chat.msgclass'>
```

```
                            <ul class = "msg_ul">
                                <li>{{chat.content.message}}</li>
                            </ul>
                        </div>
                    </div>
                </li>
            </ul>
        </div>
        <img src = "image/note.gif">{{me.user.username}}
        <input type = "text" style = "width:450px" v - model = "me.message">
        <button @click = "sendMsg">发  送</button>
    </div>
     ~
})
```

说明:

① 任何用户的数据包含两个部分: 用户 user 本身的信息, 例如用户名、头像、角色; 用户发送的聊天文字 message, 默认值为未登录提示。

② 使用 decodeURIComponent 对用户名进行解码, 避免中文用户名显示乱码问题。

③ 在页面上显示的聊天数据包含三个部分:
   ◇ divclass 样式, 这是一个层, 决定了聊天窗是显示在聊天区域的右边还是左侧;
   ◇ msgclass 样式, 这也是一个层, 用来显示具体的聊天信息 content。msgclass 决定了该层在聊天窗中是靠左还是靠右显示;
   ◇ content 是一个对象, 代表了聊天内容, 其内容包含用户对象 user、聊天文字 message。通过 divclass、msgclass 就可决定当前用户的聊天信息显示在聊天区域的右侧, 其他用户的则显示在左侧。

12.4.6 实现 JWT 令牌验证

下面来修改一下第 9 章创建的 JWTAssit 令牌工具类, 加入令牌有效性验证方法:

```java
public boolean verify(String jwt, String username) {
    boolean isLegal = false;
    try {
        JWSInput input = new JWSInput(jwt, ResteasyProviderFactory.getInstance());
        boolean b = RSAProvider.verify(input, KEY_PAIR.getPublic());
        if (b) {
            String token = input.readContent(String.class);
            JsonNode node = new ObjectMapper().readTree(token);
            long expires = node.get("exp").asLong();                              //有限期限
            String iss = node.get("iss").asText();                                //签发人
            String sub = node.get("prn").asText();                                //主题
            String aud = URLDecoder.decode(node.get("aud").asText(), "utf - 8");  //受让人
            isLegal = iss.equals("ccgg") && sub.equals("myweb") &&
```

```
                    aud.equals(URLDecoder.decode(username,"utf-8"))
                    && System.currentTimeMillis() <= expires;
            }
        } catch (Exception e) { e.printStackTrace(); }
        return isLegal;
    }
```

代码利用 JWSInput 读入令牌 JWT,并利用 RSAProvider.verify()方法进行验证。若是合法令牌,则对令牌中的内容进行比对,进一步验证。

12.4.7　配置 Kafka 和 WebSocket 全局参数

在 application.properties 中配置 Kafka:

```
#配置 Kafka
spring.kafka.bootstrap-servers=localhost:9092
spring.kafka.consumer.max-poll-records=15
spring.kafka.consumer.auto-offset-reset=latest
#配置 WebSocket 端点
chat.ws-endpoint=/schat/{username}/{token}
#配置 Kafka 主题
chat.topic-name=schat
```

Kafka 提供了丰富的配置项。这里仅分别配置了 Kafka 服务器(默认端口为 9092)、拉取消息时返回的最大记录数(15)、数据消费模式为最新数据模式(latest)。

WebSocket 端点 schat 传送了两个参数:用户名、JWT 令牌,用于在聊天窗显示当前用户名及登录有效性验证。

12.4.8　WebSocket 配置类及拦截器

```
@Configuration
@EnableWebSocket
@RequiredArgsConstructor
public class ChatConfig implements WebSocketConfigurer {
    @Value("${chat.ws-endpoint}")
    private String wsEndPoint;                              //端点
    private final @NonNull ChatService sChat;               //注入聊天服务类
    private final @NonNull JWTAssit jwtAssit;               //注入令牌工具类

    @Override
    public void registerWebSocketHandlers(WebSocketHandlerRegistry reg) {
        reg.addHandler(sChat, wsEndPoint)                   //注册聊天服务,并配置端点
            .addInterceptors(chatInterceptor())             //添加拦截器
            .withSockJS();                                  //启用 SockJS 支持
```

```java
        }
        private HandshakeInterceptor chatInterceptor() {                                    //验证用户令牌
            return new HttpSessionHandshakeInterceptor() {
                @Override
                public boolean beforeHandshake(ServerHttpRequest request,
                        ServerHttpResponse response, WebSocketHandler wsHandler,
                                    Map<String, Object> attributes) {
                    String[] uri = request.getURI().toString().split("/");     //分割SockJS地址
                    return jwtAssit.verify(uri[uri.length - 4], uri[uri.length - 5]);   ①
                }
            };
        }
    }
```

说明：

① 后端怎么拿到 SockJS 传送的用户名、令牌？根据前面端点定义格式"/schat/{username}/{token}"，以及前面介绍的 SockJS 地址规范，发送地址类似于这样：ws://localhost:8080/schat/%E6%9D%A8%E8%BF%87/eyJhbG……/356/t1xawrqg/websocket，地址用"/"分割成数组 uri 后，倒数第 4 个就是令牌，即 uri[uri.length-4]。显然，uri[uri.length-5]是用户名了。

12.4.9 创建聊天服务

1. 创建服务类 ChatService

```java
@Service
@RequiredArgsConstructor
public class ChatService extends TextWebSocketHandler {
    @Value("${chat.topic-name}")
    private String topicName;
    private final @NonNull ConsumerFactory consumerFactory;    //注入消费者工厂
    private final @NonNull KafkaProperties kafkaProp;          //注入 Kafka 属性
    private final @NonNull Environment env;                    //注入应用环境

    @Override
    public void afterConnectionEstablished(WebSocketSession session) {
        pollLatestMsg(session);
    }

    @Override
    protected void handleTextMessage(WebSocketSession session, TextMessage message) {
        SenderOptions options =
                        SenderOptions.create(kafkaProp.buildProducerProperties());
        KafkaSender.create(options).send(
            Mono.just(SenderRecord.create(
                new ProducerRecord(topicName, message.getPayload()), session.getId())
```

```
            ).subscribe();
        }                                                                    ①
}
```

说明：

① 首先创建了一个生产记录 ProducerRecord，将要发的消息作为其内容，据此创建一个发送记录 SenderRecord，然后将其包裹为发布单量 Mono。最后，利用发送器 KafkaSender 将消息推送出去。

成功连接 WebSocket 后，在 afterConnectionEstablished() 方法中调用 pollLatestMsg()，从 Kafka 拉取最近 15 条聊天记录显示给当前用户。

2. pollLatestMsg()方法

要拉取最近 15 条记录，需要用 Kafka 提供的 endOffsets() 获取最后一条记录的偏移量，简单计算后返回开始拉取位置的偏移量，然后通过 seek(offset) 方法定位到倒数第 15 条记录（若总共不足 15 条则从 0 开始），最后就可从该记录位置开始，拉取出到末尾的全部记录。

```java
private void pollLatestMsg(WebSocketSession session) {
    long maxPoll = Long.parseLong(
env.getProperty("spring.kafka.consumer.max-poll-records"));   //读取环境参数值
    TopicPartition tp = new TopicPartition(topicName, 0);          //Kafka 主题分区
    long endOffset = (long) consumerFactory.createConsumer()
            .endOffsets(Collections.singletonList(tp)).getOrDefault(tp, 0);
                                                                   //最末偏移量
    long offset = endOffset > maxPoll ? endOffset - maxPoll : 0;   //计算出开始拉取位置

    ReceiverOptions<Integer, String> options =
        ReceiverOptions.<Integer, String>create(kafkaProp.buildConsumerProperties())
        .consumerProperty(ConsumerConfig.GROUP_ID_CONFIG, session.getId() + "-group")
        .addAssignListener(partitions -> partitions.forEach(p -> p.seek(offset)))
                                                                   //到指定位置
        .commitBatchSize(5)                                        //批量处理数
        .subscription(Collections.singleton(topicName));           //订阅该主题
    KafkaReceiver.create(options).receive()                        //接收消息
        .subscribe(r -> {
            try {
                session.sendMessage(new TextMessage(r.value()));   //发送出去
            } catch (IOException e) { e.printStackTrace(); }
        });
}
```

12.5 场景任务挑战——学生、教师各自的聊天室

上一节的聊天室，并没有区分用户角色问题。现在，请读者实现：同一角色用户可以互相聊天，不同角色的用户之间不能聊天。

第 13 章

教务辅助管理项目开发

教务辅助管理系统(Teaching Assistant Management System,TAMS)用于学生、教师日常教学辅助管理。本章主要介绍 TAMS 的主要功能、技术选型、具体功能实现。本章既是前面所学知识的贯通运用,又是技术方法的提升。

13.1 系统概述

教务辅助管理系统是一个基于 B/S 模式,用于学生、教师一些常规事项管理的项目。为了方便理解,这里对系统进行了功能调整、简化处理。

13.2 系统功能简介

系统主界面如图 13-1 所示,主要包括以下 9 个部分。
- 首页:显示主界面及导航菜单。
- 消息推送:后台向登录的用户推送通知、公告等短消息。用户登录后,只会接收当天推送的消息。
- 用户登录:登录成功后,会生成 JWT 令牌以备其他功能验证使用。
- 用户注册:根据用户名、密码、E-mail、用户角色进行注册。用户名唯一,不能重复注册。与前面章节不同,这里对 Users 实体类进行了持久化处理。
- 学院风采:展示各学院的文字介绍及视频介绍。先展示文字内容,停顿 10s 后,自动播放视频介绍。这里只实现了会计学院的文字、视频处理。由于其他学院处理方法类似,故略去。各学院的视频文件,使用 Spring WebFlux 的流媒体方式加载,这样一来视频播放就无须等待整个视频内容下载完成即可进行。
- 学生查询:只有登录的用户才能查询。查询时,可输入姓名或班级关键字,进行模糊查询。另外,这里的前端没有使用 template 模板,而是全部用 render()渲染实现。

图 13-1 主界面

◇ 招生一览：可查询指定学院某专业近五年的招生数据情况，用 ECharts 饼图显示招生数据。这里通过学院、专业下拉选择框的动态关联，实现不同学院、不同专业的招生饼图展示。
◇ 资料上传：登录过的用户才能上传文件。上传文件大小限制为 50MB，使用 WebFlux 文件数据流传送文件。上传后的文件名格式为：上传时间-用户名-文件名，例如：20210901213056-杨过-奖学金一览表.xls。
◇ 交流空间：登录用户的讨论、学习场所。虽是在第 12 章基础上进行的改造，但有很大不同：使用 Spring WebFlux 实现群聊处理。

13.3 系统技术选型

前端使用 Vue 3，后端基于 Spring WebFlux。数据库仍然是 PostgreSQL，采用 R2DBC 连接数据库。服务器稍微变化了一下，使用 WebFlux 默认的、广受欢迎的 Netty。

13.3.1 前端组件化

前端文件结构，如图 13-2 所示。
由图 13-2 可知，整个系统相应的功能定义成以下 6 个组件。
◇ chat.component.js：登录用户交流讨论的组件。
◇ enroll.component.js：招生组件，例如招生数据一览。
◇ file-upload.component.js：文件上传组件。
◇ school.component.js：学院管理组件，例如学院风采。
◇ student-query.component.js：学生信息查询组件。
◇ users.component.js：用户管理组件，例如用户登录、注册处理。

图 13-2 前端文件结构

这些 JS 文件放置在 module 文件夹下。同时，将系统所需的 JS 文件，例如 Vue.js、MD5、Axios、ECharts 等，放在 require 文件夹下。图中的 modules.aggregator.js 是组件聚合器，用于集中组件的导入导出管理。

13.3.2　后端模块化

整个后端文件的包结构如图 13-3 所示。

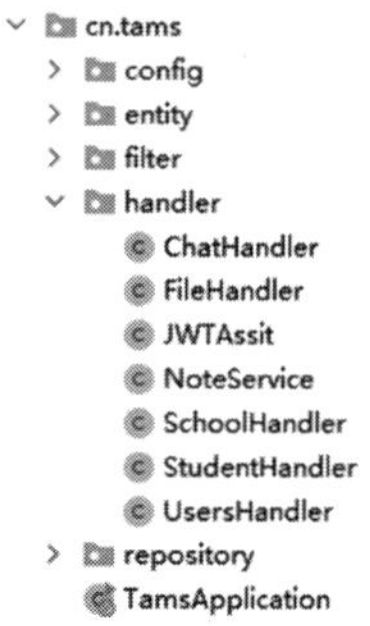

系统后端处理定义成以下 7 个功能模块。

◇ ChatHandler：登录用户讨论交流聊天的组件类。
◇ FileHandler：处理文件上传的组件类。
◇ JWTAssit：处理 JWT 令牌生成与验证的组件类，已在第 12 章全部实现。
◇ NoteService：消息推送的服务类。
◇ SchoolHandler：处理学院风采视频文件的组件类。

图 13-3　后端的包结构

◇ StudentHandler：处理与学生业务相关的组件类，例如学生信息模糊查询、招生数据等。
◇ UsersHandler：处理与用户业务相关的组件类，例如用户登录、用户注册。

项目包结构的说明：

◇ config：配置包，用于消息服务配置、请求路由配置、WebFlux 全局配置、聊天端点配置。
◇ entity：实体包，存放实体类文件。
◇ filter：过滤包，用于请求访问的用户身份验证过滤。
◇ repository：数据访问包，存放扩展自 R2dbcRepository 的数据访问接口类。

13.4　数据表设计

前面章节已经陆续将一些表创建好了，分别是用户表 users、学生表 student、学院表 school、专业表 major、招生数据表 enroll。与前面章节相比，这里只多了一个消息表 note。note 表用于存放发布的通知、公告等消息内容，表结构如图 13-4 所示。

图 13-4　note 表结构

这几个字段含义分别是：id，消息记录号，主键由数据库自动生成；content，消息内容；issuer，消息发布者；issuetime，发布时间。

13.5 系统实现

为了节省篇幅，实现过程中的 CSS 代码全部略去。

13.5.1 创建 Spring Reactive Web 项目

依次选择菜单 File→New→Project→Spring Initializr，输入项目名称 tams 及其他信息定义，如图 13-5 所示。注意，Packaging 选择 Jar。再单击 Next 按钮，选择 Web 下的 Spring Reactive Web，如图 13-6 所示。然后单击 Finish 按钮完成 tams 项目创建。

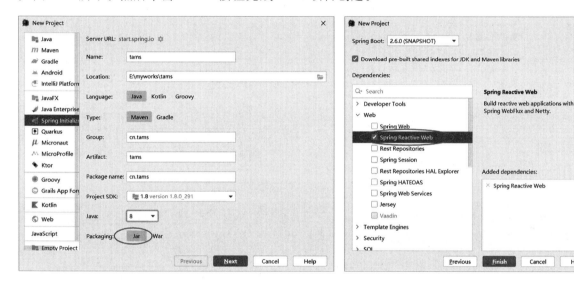

图 13-5　创建 tams 项目　　　　　图 13-6　选择 Spring Reactive Web

最后，请按图 13-2 和图 13-3 所示，分别创建好 resources/static 下的文件夹结构、cn.tams 下的包结构。

13.5.2 配置 application.yml 全局参数

在 application.yml 中输入应用的各种参数：

```
#服务器配置
server:
    address: 127.0.0.1
    port: 80
#请求路由的地址配置
tams:
    url:
```

```yaml
    home: classpath:/static/home.html
    note: /note/event,/note/get
    usr: /usr,/login,/regist
    stu: /stu,/query/{key}/{username},/query/sm,/enroll/{mno}
    file: /file/up/{username}
    upload-path: tamsfiles
    school-video: /video/{name}
    school-video-file: classpath:/static/video/%s.mp4
#数据库、文件上传、Kafka配置
spring:
  r2dbc:
    url: r2dbc:postgresql://${server.address}:5432/tamsdb
    username: admin
    password: '007'
    pool:
      enabled: true
      initial-size: 8
      max-size: 50
      max-idle-time: 1500000 //25分钟
  servlet:
    multipart:
      max-file-size: 50MB
      max-request-size: 50MB
  kafka:
    bootstrap-servers: ${server.address}:9092
    consumer:
      max-poll-records: 15
      auto-offset-reset: latest
#聊天配置
chat:
  ws-endpoint: /schat/{username}/{token}
  topic-name: schat
  logoutMsg: "您的登录身份已过期失效,无法向他人发送消息,请重新登录!"
```

13.5.3　加入项目主要依赖

在 pom.xml 中新增以下依赖项:

```xml
<dependency>
    <groupId>io.r2dbc</groupId>
    <artifactId>r2dbc-postgresql</artifactId>
    <version>0.8.8.RELEASE</version>
</dependency>
<dependency>
    <groupId>org.springframework.boot</groupId>
    <artifactId>spring-boot-starter-data-r2dbc</artifactId>
</dependency>
```

```xml
<dependency>
    <groupId>org.springframework.kafka</groupId>
    <artifactId>spring-kafka</artifactId>
</dependency>
<dependency>
    <groupId>org.projectlombok</groupId>
    <artifactId>lombok</artifactId>
    <version>1.18.20</version>
    <scope>provided</scope>
</dependency>
<dependency>
    <groupId>org.jboss.resteasy</groupId>
    <artifactId>resteasy-core</artifactId>
    <version>4.6.1.Final</version>
</dependency>
<dependency>
    <groupId>org.jboss.resteasy</groupId>
    <artifactId>jose-jwt</artifactId>
    <version>4.6.1.Final</version>
</dependency>
```

包括原有的 spring-boot-starter-webflux，总共 7 个主要依赖项。

13.5.4 引入 JS 支持文件

将 Vue、Axios、MD5、ECharts 等 JS 支持文件复制到项目 resources/static/require 文件夹下，如图 13-7 所示。

13.5.5 使用聚合器管理组件

在 resources/static 下创建聚合器 modules.aggregator.js，管理组件的导入导出。modules.aggregator.js 主要包括以下 3 部分。

图 13-7 JS 支持文件

1）导入 module 文件夹下的 6 个组件

```js
import {UsersLoginComponent as login, UsersRegistComponent as regist} from "./module/users.component.js"
import {SchoolsComponent as school} from "./module/school.component.js"
import {StudentQueryComponent as query} from "./module/student-query.component.js"
import {EnrollComponent as enroll} from "./module/enroll.component.js"
import {FileUploadComponent as upload} from "./module/file-upload.component.js"
import {ChatComponent as chat} from "./module/chat.component.js"
```

2）定义系统菜单项

```js
export const menus = [
    {id: 'home', name: '首页', module: null},
    {id: 'login', name: '用户登录', module: login},
```

```
    {id: 'regist', name: '用户注册', module: regist},
    {id: 'school', name: '学院风采', module: school},
    {id: 'query', name: '学生查询', module: query},
    {id: 'enroll', name: '招生一览', module: enroll},
    {id: 'upload', name: '资料上传', module: upload},
    {id: 'chat', name: '交流空间', module: chat}
]
```

3) 定义系统公共属性、函数

```
window.tamsPub = {
    curModule: Vue.ref(null),        //当前组件
    loginer: Vue.reactive({
        username: null,              //用户名
        logo: null,                  //用户图标
        msg: null,                   //后台推送的消息内容
        msgnum: null                 //后台推送的消息数
    }),
    setLoginer(data) {
        this.loginer.username = data.username
        this.loginer.logo = data.logo
        this.loginer.msgnum = 0
        sessionStorage.setItem('loginer', JSON.stringify(data))
    },
    clearLoginer() {
        Object.keys(this.loginer).forEach(k => this.loginer[k] = null)
        sessionStorage.removeItem('loginer')
    }
}
```

在这里定义了一个浏览器窗口对象 tamsPub,该对象属于全局对象,只需要在 Vue 主应用中返回,其他组件无须导入就可直接使用。该对象包含:

（1）curModule 属性,代表主页导航菜单项切换时,当前需要挂载的组件；

（2）loginer 对象,代表当前登录用户；

（3）setLoginer()函数,用来设置登录成功用户的用户名、logo 图像、当前接收到的消息数,并将当前登录用户信息以 sessionStorage 方式保存到 loginer 中；

（4）clearLoginer()函数,清除登录状态,即将 loginer 对象的全部属性清空,并清除 sessionStorage 内容。

13.5.6 应用入口程序

```
@SpringBootApplication
public class TamsApplication {
    public static void main(String[] args) {
```

```
            SpringApplication.run(TamsApplication.class, args);
    }
}
```

请大家比较一下与第 9 章入口程序的差异。

13.5.7　WebFlux 配置和路由配置

1. WebFlux 配置

在 cn.tams.config 包下新建 WebConfig 配置类：

```
@Configuration
@EnableWebFlux
public class WebConfig implements WebFluxConfigurer {

    @Override
    public void addResourceHandlers(ResourceHandlerRegistry registry) {
        registry.addResourceHandler("/**")
                .addResourceLocations("classpath:/static/")
                .setCacheControl(CacheControl.maxAge(8, TimeUnit.HOURS)
                        .cachePublic());
    }

}
```

WebConfig 类启用 WebFlux，需要改写 addResourceHandlers()方法：注册静态资源所在的路径为 static 文件夹。静态资源是指图片、HTML 文件、JS 文件等。注册后，应用就能够访问到这些静态资源文件。另外，启用了对静态资源的缓存处理，缓存时间为 8h，以便提高访问速度。

2. 请求路径的路由配置

在 cn.tams.config 包下新建 RouteConfig 配置类：

```
@Configuration(proxyBeanMethods = false)                //不使用代理 Bean 模式
@RequiredArgsConstructor
public class RouteConfig {
    private static final RequestPredicate ACCEPT_JSON =
                                    accept(MediaType.APPLICATION_JSON);
    private static final RequestPredicate FORM_DATA =
                                    accept(MediaType.MULTIPART_FORM_DATA);
    @Value("${tams.url.home}")                          //在 application.yml 中定义
    private Resource home;                              //主页文件位置
    @Value("${tams.url.note}")
    private String[] noteEndPoint;                      //消息推送请求路径
    @Value("${tams.url.usr}")
    private String[] usrEndPoint;                       //用户登录请求路径
```

```java
@Value("${tams.url.school-video}")
private String videoUrl;                                     //学院视频文件的请求路径
@Value("${tams.url.stu}")
private String[] stuEndPoint;                                //学生信息处理的请求路径
@Value("${tams.url.file}")
private String fileEndPoint;                                 //文件处理的请求路径
private final @NonNull AuthFilter authFilter;                //注入用户身份验证



//项目的路由地址:http://localhost/,对应打开 static/home.html 文件
@Bean
public RouterFunction<ServerResponse> homeRoute() {
    return route(GET("/"), request -> ok().bodyValue(home));
}
//前端消息事件连接端点地址:/note/event;后端获取消息记录地址:/note/get
@Bean
public RouterFunction<ServerResponse> noteRoute(NoteService handler) {
    return route()
            .GET(noteEndPoint[0], handler::eventNotes)       //消息事件
            .GET(noteEndPoint[1], handler::getNotes)         //获取数据库消息记录
            .build();
}
//用户管理的路由地址.用户登录:/usr/login;用户注册:/usr/regist
@Bean
public RouterFunction<ServerResponse> usrRoute(UsersHandler handler) {
    return route()
            .path(usrEndPoint[0], r -> r                     //登录请求的根路径,即/usr
                    .POST(usrEndPoint[1], ACCEPT_JSON, handler::login)
                    .POST(usrEndPoint[2], ACCEPT_JSON, handler::regist))
            .build();
}
//学院风采管理中视频文件的路由地址:/video/{name},路径参数 name 表示不包含
//扩展名的视频文件名,例如/video/acc 表示请求的视频文件为:/video/acc.mp4
@Bean
public RouterFunction<ServerResponse> schoolRouter(SchoolHandler handler) {
    return route()
            .GET(videoUrl, handler::videoHandler)
            .build();
}
//学生管理的路由地址.
//关键字查询:/stu/query/{key}/{username};招生数据一览:/stu/enroll/{mno};
//获取全部学院名称、专业名称:/stu/query/sm
@Bean
public RouterFunction<ServerResponse> stuRoute(StudentHandler handler) {
    return route()
            .path(stuEndPoint[0], r -> r                     //请求的根路径,即/stu
                    .GET(stuEndPoint[1], handler::queryStuKey) //关键字查询
```

```
                        .filter(authFilter))                          //身份验证
                    .path(stuEndPoint[0], r -> r                      //请求的根路径,即/stu
                        .GET(stuEndPoint[2], handler::allSchoolMajor)  //学院专业
                        .GET(stuEndPoint[3], handler::enroll5Mno))    //招生数据
                    .build();
    }
    //文件管理的路由地址.文件上传:/file/up/{username}
    @Bean
    public RouterFunction<ServerResponse> fileRoute(FileHandler handler) {
        return route()
                    .POST(fileEndPoint, FORM_DATA, handler::uploadFile)  //文件上传
                    .filter(authFilter)                               //身份验证
                    .build();
    }
}
```

在路由地址中,usrRoute、stuRoute 都对路径进行了分类设置:/usr、/stu,再在下面划分子类地址,例如/usr/login、/usr/regist 等。另外,对学生信息查询、文件上传,添加了身份验证过滤处理。

13.5.8 身份验证过滤组件

在 cn.tams.filter 包下创建对用户身份进行有效性验证的 AuthFilter 类:

```
@Component
@RequiredArgsConstructor
public class AuthFilter implements HandlerFilterFunction {
        private final @NonNull JWTAssit jwtAssit;         //注入JWTAssit令牌组件类

        @Override
        public Mono filter(ServerRequest request, HandlerFunction next) {
            String username = request.pathVariable("username");                        ①
            String token = request.headers().firstHeader(HttpHeaders.AUTHORIZATION);   ②
            if (jwtAssit.verify(token, username))
                    return next.handle(request);
            else
                    return ServerResponse
                                .status(HttpStatus.UNAUTHORIZED)
                                .build(Mono.empty());
        }
}
```

说明:

① 获取请求路径中 username 参数的值,这里指的是学生信息查询请求路径"/stu/query/{key}/{username}"、文件上传请求路径"/file/up/{username}"中的路径参数{username}。

② 获取前端提交的请求数据中 HTTP 头 Authorization 中存放的 JWT 令牌。

13.5.9 主页

在resources/static下创建home.html、home.js。为简便起见,主页中引入CSS文件夹下的样式文件及require下的JS支持文件的相关代码全部略去。主页效果如图13-1所示。

1. home.html

```html
<div id="tamsApp">
    <div class="home-title">
        教务辅助管理系统
        <div class="home-loginer" v-show="tamsPub.loginer.username!=null">
            <img :src="tamsPub.loginer.logo" height="16" width="16">
            {{tamsPub.loginer.username}}
            <div class="home-msg" :title="tamsPub.loginer.msg">        ①
                {{tamsPub.loginer.msgnum}}
            </div>
        </div>
    </div>
    <div class="home-link">
        <ul class="home-ul">
            <li @click="menuIndex = index" v-for="(item,index) in menus">   ②
                {{item.name}}
            </li>
        </ul>
    </div>
    <div class="home-detail">
        <transition mode="out-in">
            <component :is="tamsPub.curModule.value"></component>
        </transition>
    </div>
</div>
<script src="home.js" type="module"></script>
```

说明:

① 绑定<div>的title属性显示后台服务器推送的消息,并显示消息条数。

② menuIndex代表导航菜单索引。单击导航菜单时,menuIndex值的变化,就会激发Vue的watch侦听(见下面home.js中的watch),进而改变curModule的值。<component>使用的还是我们前面用过的动态组件,通过is属性动态加载curModule所代表的组件。<transition>定义了组件切换动画,相关CSS定义如下:

```css
.v-enter, .v-leave-to {
    opacity: 0;
    transform: scaleX(0.1);
}
.v-enter-active, .v-leave-active {
```

```
        transition: all 0.6s ease;
}
```

使用 v-这种样式名称，会自动绑定到 transition，无须显示指定。scaleX 横向缩小组件界面至 0.1，最后消失。

2. home.js

```
import {menus} from "./modules.aggregator.js"     //从聚合器中导入组件
const home = {
        setup() {
                const {ref, watch} = Vue
                const menuIndex = ref(null)                //菜单索引
                watch(menuIndex, (newIndex) => tamsPub.curModule.value = menus[newIndex].id)
                return {menus, tamsPub, menuIndex}
        }
}
const tamsApp = Vue.createApp(home)
menus.forEach((menu) => {
        tamsApp.component(menu.id, menu.module)    //注册菜单组件
})
tamsApp.mount('#tamsApp')
```

通过侦听 menuIndex 索引值的变化，改变 tamsPub 对象 curModule 属性的值，从而触发动态加载该菜单索引对应组件的目的。

13.5.10 用户登录

用户登录界面如图 13-8 所示。

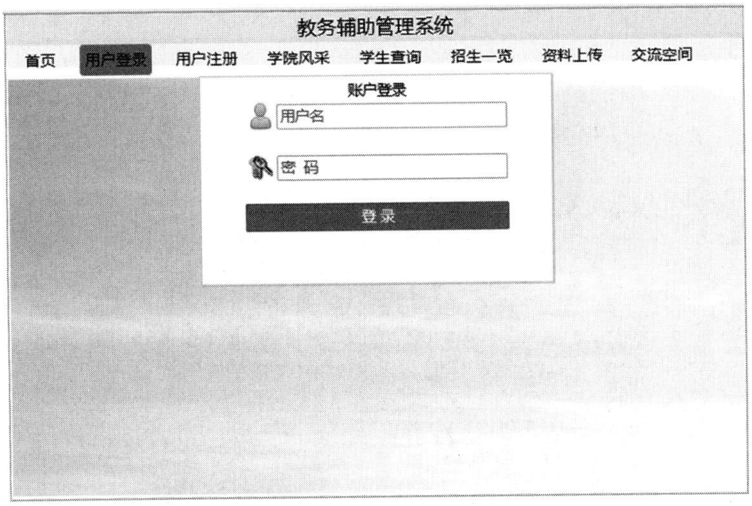

图 13-8　用户登录界面

1. 前端：users.component.js

在 resources/static/module 下创建 users.component.js 组件：

```js
export const UsersLoginComponent = Vue.defineComponent({
    setup() {
        const user = Vue.reactive({
            username: '',
            password: '',
            isDoing: false,        //登录进行状态标识
            state: null,           //登录结果状态
        })
        const {doLogin} = UsersMethods                                          ①
        return {user, doLogin}
    },
    template:
        `<div class = "user-login">
            账户登录<br>
            <img src = "image/username.png">
            <input type = "text" class = "login-input" placeholder = "用户名"
                                                       v-model = "user.username">
            <br><br><img src = "image/password.png">
            <input type = "password" class = "login-input" placeholder = "密　码"
                                                   v-model = "user.password"><br><br>
            <button class = "login-button" @click = "doLogin(user)">登　录</button><br>
            <div v-show = "isDoing">
                <img src = "image/progress.gif" width = "50" height = "50"/>
                                                         正在登录,请稍候……
            </div>
            <span style = "color:#ff0000">{{user.state}}</span>
        </div>`
})
```

说明：

① 与第 12 章的代码不同，这里定义了一个用户函数集 UsersMethods，从中解构出 doLogin() 函数。函数集 UsersMethods 代码如下：

```js
const UsersMethods = (function (exports) {
    const doLogin = async (user) => {              //执行登录
        user.isDoing = true                         //正在进行登录
        user.state = ''                             //登录结果默认为空
        tamsPub.clearLoginer()                      //重置聚合器中的 loginer 对象属性为空
        user.password = CryptoJS.MD5(user.password) + ''
        try {
            const {data, status} = await axios({
                method: 'post',
                url: 'usr/login',
                data: user
```

```
                })
                if (data == null)                          //登录失败
                    return
                if (data.token != null && status == '200') {  //登录成功
                    tamsPub.setLoginer(data)               //设置聚合器中loginer对象的值
                    tamsPub.curModule.value = null         //关闭登录界面
                    getServerNotes()                       //获取当天服务器推送的消息
                } else
                    user.state = '用户名或密码错误,登录失败!'
            } catch (error) {
                user.state = '系统连接错误,无法登录!'
            }
            user.isDoing = false
        }
        //获取当天后台服务器推送的消息
        const getServerNotes = () => {
            let es = new EventSource('note/event');         //开启一个持久化的连接事件
            let notes = ''                                  //默认消息为空
            es.onmessage = function (event) {               //接收到消息
                let jd = JSON.parse(event.data)             //将消息数据解析为JSON对象
                if (!notes.match(jd.content))               //是新消息
                    notes += jd.content + '\u3000' + jd.issuer + ' ' + jd.issuetime + '\n'  ①
                let regex = new RegExp(/\n/g);                                              ②
                let result = notes.match(regex);
                tamsPub.loginer.msgnum = result ? result.length : 0;
                tamsPub.loginer.msg = notes           //消息赋值给msg属性以便index.html中显示
            }
        }
        exports.doLogin = doLogin
        return exports;
}({}))
```

说明:

① \u3000 为全角空格,\n 表示换行。这里构建了多个消息连接而成的字符串。

② 我们需要知道接收了多少条消息,以便在主界面的红圈圆点内显示新消息数量。消息已经以字符串形式保存在 notes 中,利用正则表达式,通过统计换行符\n 的个数,就可知道有几条消息。先定义规则:/\n/g,该规则由"/"和"\n/g"组成,表示全局匹配换行\n,再利用 match 在 notes 内匹配,最后利用 result.length 获得消息的条数,赋值给 loginer 对象的 msgnum 属性。

2. 后端:Users 实体类

与第 9 章 9.5.2 节代码相同。

3. 后端:UsersRepository 数据访问接口类

```
@Repository
public interface UsersRepository extends R2dbcRepository<Users, String> {}
```

只是定义了一个接口类,并没有自定义一些方法。

4. 后端:UsersHandler 业务处理类

在 cn.tams.handler 包下新建 UsersHandler 类:

```java
@Component
@RequiredArgsConstructor
public class UsersHandler {
    private final @NonNull UsersRepository repository;
    private final @NonNull JWTAssit jwtAssit;

    public Mono<ServerResponse> login(ServerRequest request) {
        return request.bodyToMono(Users.class)
                .flatMap(u ->
                        repository.findOne(Example.of(u))                       //登录成功
                                .flatMap(usr -> {
                                    Map<String, String> map = new HashMap(4);
                                    map.put("username", usr.getUsername());
                                    map.put("logo", usr.getLogo());
                                    map.put("role", usr.getRole());
                                    map.put("token", jwtAssit.getToken(usr.getUsername()));
                                    return ok().bodyValue(map);
                                })
                                .switchIfEmpty(ServerResponse.noContent().build())  //登录失败
                                .onErrorResume(e -> ok().bodyValue(e.getMessage()))  //出错
                );
    }
}
```

在以前的章节中,生成的令牌是通过 HTTP 头 Authorization 向前端传送的。这里改变了一下处理方法:全部放在 map 中返回。读者可与第 12 章进行比较。一个问题常可从多角度去思考、解决。

13.5.11 消息推送

用户登录成功后,就可请求接收后端推送的当天的消息。接收到消息后,在用户名的右上角会用红色小圆圈显示消息数。鼠标指向红色圆圈,就会显示消息的具体内容,如图 13-9 所示。

图 13-9 接收的消息

前面用户登录成功后,就会调用 getServerNotes() 函数准备接收后端推送的消息。现在只需要编写后端消息推送程序就可以了。

1. 消息实体类 Note

在 cn.tams.entity 包下创建 Note 类：

```java
@Data
@NoArgsConstructor
@AllArgsConstructor
public class Note implements Serializable {
    @Id
    private int id;                    //记录号
    private String content;            //内容
    private String issuer;             //发布者
    private String issuetime;          //发布时间
}
```

2. 消息记录访问接口类 NoteRepository

在 cn.tams.repository 包下新建 NoteRepository 类：

```java
@Repository
public interface NoteRepository extends R2dbcRepository<Note, String> {
    String query = "select content,issuer,\"time\"(issuetime)::text as issuetime " +
                   "from note where date(issuetime) = current_date order by id desc";
    @Query(query)
    Flux<Note> getTodayNotes();
}
```

3. 消息推送配置类 NoteConfig

在 cn.tams.config 包下新建 NoteConfig 类：

```java
@Configuration
public class NoteConfig {
    @Value("${server.address}")
    private String serverAddr;

    @Bean
    public WebClient webClient() {
        return WebClient.builder().baseUrl(serverAddr).build();
    }

    @Bean
    public Sinks.Many<Note> broadcast() {
        return Sinks.many().multicast().directBestEffort();
    }

    @Bean
    public Flux<Note> sinkFlux(Sinks.Many<Note> sink) {
        return sink.asFlux();
    }
```

```
        }
    }
```

NoteConfig 类配置了三个 Bean,其实例将在 NoteService 类中自动注入。

(1) webClient():构建了一个以"/note/get"为基准请求路径的 WebClient 对象,以便通过其获取数据库中的消息记录;

(2) broadcast:当收到来自 WebClient 的新消息时,将其传递给众多的订阅者(登录用户);

(3) sinkFlux:返回一个消息通量,并以消息发布者身份,通过事件流将新消息传递给浏览器。

4. 消息服务类 NoteService

在 cn.tams.handler 包下创建 NoteService 类:

```java
@Service
@EnableScheduling
@RequiredArgsConstructor
public class NoteService {
    @Value("${tams.url.note}")
    private String[] noteEndPoint;

    private final @NonNull NoteRepository nRepository;      //注入 NoteRepository 实例
    private final @NonNull WebClient webClient;             //注入 WebClient 实例
    private final @NonNull Sinks.Many<Note> sink;           //注入 Sinks.Many 实例
    private final @NonNull Flux<Note> notes;                //注入 Flux 实例
    //以固定的每 2 分钟一次的频率,定期向"/note/get"发出消息请求
    @Scheduled(fixedRate = 60000 * 2)
    public void publishNotes() {
        webClient.get().uri(noteEndPoint[1]).retrieve()
            .bodyToFlux(Note.class)
            .subscribe(sink::tryEmitNext);
    }
    //路由地址"/note/event"将调用此方法
    public Mono<ServerResponse> eventNotes(ServerRequest request) {
        return ok().contentType(MediaType.TEXT_EVENT_STREAM)
            .body(notes, Note.class);
    }
    //去数据库获取具体消息,即 webClient 真正调用的获取消息的方法.
    public Mono<ServerResponse> getNotes(ServerRequest request) {
        return ok().body(nRepository.getTodayNotes(), Note.class);
    }
}
```

eventNotes()方法连接注入的 Flux 实例 notes,客户端浏览器就可通过 notes 订阅消息,并将收到的消息发送给浏览器。其中,TEXT_EVENT_STREAM 表示返回给前端的数据类型为文本事件流。

这里其实还缺少一个新消息发布模块,即由管理员向 note 数据表发布新的消息内容。读者可自行补充完成该模块。

13.5.12 用户注册

用户注册界面如图 13-10 所示。

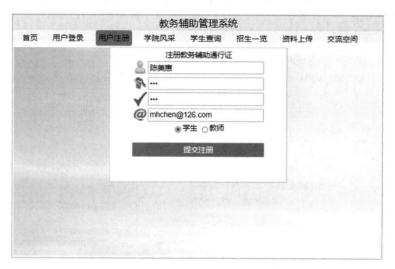

图 13-10 用户注册

1. 前端：users.component.js

在 users.component.js 文件中加入用户注册组件 UsersRegistComponent：

```
export const UsersRegistComponent = Vue.defineComponent({
    setup() {
        const {doRegist} = UsersMethods
        const user = Vue.reactive({
            username: '',
            password: '',
            repassword: '',
            role: 'student',
            isDoing: false,
            state: null,
        })
        return {user, doRegist}
    },
    template: `
        <div class="user-regist">
            注册教务辅助通行证<br>
            <img src="image/username.png">
            <input type="text" class="regist-input" placeholder="用户名" required
                                 v-model.trim="user.username"><br>
            <img src="image/password.png">
            <input type="password" class="regist-input" placeholder="密　码"
                                 required v-model.trim="user.password"><br>
            <img src="image/repwd.png">
```

```
                <input type = "password" class = "regist-input" placeholder = "确认密码"
                                            v-model = "user.repassword"><br>
                <img src = "image/email.png">
                <input type = "email" class = "regist-input" placeholder = "email"
                                            v-model = "user.email"><br>
                <input type = "radio" name = "role" value = "student" v-model = "user.role">学生
                <input type = "radio" name = "role" value = "teacher" v-model = "user.role">教师
                <br><br><button class = "regist-button" @click = "doRegist(user)">
                                            提交注册</button><br>
                <div v-show = "isDoing">
                    <img src = "image/progress.gif" width = "50" height = "50"/>
                                            正在注册,请稍候……
                </div>
                <span style = "color:#ff0000">{{user.state}}</span>
        </div>
})
```

然后,在函数集 UsersMethods 中加入 doRegist()函数:

```
const doRegist = async (user) => {
    user.state = ''
        if (user.username == '' || user.password == '') {
            user.state = '用户名或密码不能为空!'
            return
        }
        if (user.repassword != user.password) {
            user.state = '两次输入的密码不一致,请重新输入!'
            return
        }
        user.isDoing = true
        user.password = CryptoJS.MD5(user.password) + ''   //对密码进行加密
        try {
            const {status} = await axios({
                method: 'post',
                url: 'usr/regist',
                data: user
            })
            if (status == '205')                              //已被注册,需要重置注册内容
                user.state = '该用户已被注册,请更改用户名!'
            else if (status == '200')
                user.state = '用户注册成功!'
        } catch (error) {
            user.state = '出错,注册失败!'
        }
        user.isDoing = false
}
exports.doRegist = doRegist
```

这里通过后端返回的 HTTP 状态码,对注册结果进行判断。

2. 后端：Users 实体类

将用户对象保存入库时调用的是 R2dbcRepository 接口提供的 save() 方法。该方法会根据主键 username 是否为空，来决定是插入新记录还是修改记录。如果 username 为空，则认为是新记录，执行 insert 处理；否则，认为是需要修改（update）记录。当用户注册时，username 显然不为空，save() 方法会尝试去修改数据库中对应 username 的记录，显然数据库中并不存在这样的记录，导致注册失败。所以，需要修改 Users，在其中加入一个 newUser 属性。在用户注册时，明确指示 save() 方法进行 insert 操作。

```
@Data
@Table
@NoArgsConstructor
@AllArgsConstructor
public class Users implements Persistable {        //实现持久化接口
    @Id
    private String username;
    private String password;
    private String logo;
    private String role;
    private String email;

    @Transient                                      //声明该属性不属于数据库，而是临时使用
    private boolean newUser;

    public Users setAsNew() {                       //设置为 true，表示需要 insert 而非 update
        newUser = true;
        return this;
    }

    @Override
    public String getId() {                         //返回主键值
        return username;
    }

    @Override
    public boolean isNew() {                        //是 insert 还是 update
        return newUser || username == null;
    }
}
```

3. 后端：UsersRepository 数据访问接口类

与前面用户登录中的一样。

4. 后端：UsersHandler 业务处理类

在 UsersHandler 类中添加 regist() 方法：

```
@Transactional                                                          //启用事务处理
public Mono<ServerResponse> regist(ServerRequest request) {
    return request.bodyToMono(Users.class)
```

```
        .flatMap(user ->
          repository.findById(user.getUsername())
            .flatMap(u -> ServerResponse.status(HttpStatus.RESET_CONTENT).build())    ①
            .switchIfEmpty(ok().build(repository.save(user.setAsNew()).then()))   //保存入库
            .onErrorResume(e -> ok().bodyValue(e.getMessage())))
        );
}
```

说明：

① 用户名已存在，需重置注册内容，所以返回一个 HTTP 状态码 205 来告知前端。

13.5.13　学院风采

学院风采画面效果如图 13-11 和图 13-12 所示。

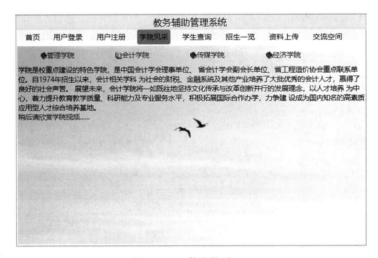

图 13-11　学院风采（1）

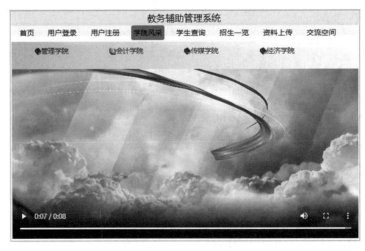

图 13-12　学院风采（2）

1. 前端：school.component.js

在 resources/static/module 下创建 school.component.js：

```js
export const SchoolsComponent = Vue.defineComponent({
    setup() {
        const schools = [
            {id: 'school-mg', name: '管理学院', module: MgComponent},
            {id: 'school-ac', name: '会计学院', module: AcComponent},
            {id: 'school-md', name: '传媒学院', module: MdComponent},
            {id: 'school-ec', name: '经济学院', module: EcComponent}
        ]
        const tamsApp = document.getElementById("tamsApp").__vue_app__      ①
        schools.forEach(item => {          //遍历 schools 数组中的各学院组件定义
            new Promise(() => {            //异步注册组件
                tamsApp.component(item.id, item.module)
            })
        })
        const {ref, watch} = Vue
        const curSchool = ref(schools[0].id)
        const sIndex = ref(0)
        watch(sIndex, () => curSchool.value = schools[sIndex.value].id)
        const setImage = (index) => sIndex.value == index
                          ? 'image/bookopen.gif' : 'image/bookclose.gif'    ②
        return {schools, curSchool, sIndex, setImage}
    },
    template: `
        <div id="school-intro">
            <div class="school-ul">
                <ul class="school-li">
                    <li v-for="(school,index) in schools" @click="sIndex = index">
                        <img :src="setImage(index)">{{school.name}}
                    </li>
                </ul>
            </div>
            <div class="school-detail">
                <keep-alive>
                    <component :is="curSchool"></component>
                </keep-alive>
            </div>
        </div>
    `
})
//会计学院文字介绍信息
const acDescribe = {
    template: `
        <div>
            <span style='float:left'>学院是校重点建设的特色学院……(略去)</span>
```

```
                    <span style='float:left;color:#0000ff'>稍后请欣赏学院视频……</span>
            </div>`
}
//会计学院视频信息
const AcComponent = Vue.defineAsyncComponent({
    loader: () => new Promise(resolve => {
        setTimeout(() => {                   //利用setTimeout函数停顿10秒钟,以便用户阅读介绍文字
            resolve({
                template:
                    `<video width = "756" height = "380" src = "video/acc"                    ③
                                    preload = "none" controls autoplay = "autoplay"/>`
            })
        }, 10000)
    }),
    delay: 0,                            //立即加载
    loadingComponent: acDescribe,        //先加载学院文字介绍内容
    errorComponent: Vue.h('span', "视频加载失败……")
})
…                                         //其他学院略去
```

① 需要在 school.component.js 组件里面给 home.js 中定义的 tamsApp 注册各个学院组件,由于 tamsApp 并不是全局对象,这里并不能直接使用! 有两种解决方法: 第一种,将 home.js 中的 tamsApp 定义在 window 对象上; 第二种,利用 Vue 创建应用后生成的__vue_app__,这个实际上代表了 tamsApp。这里采用了第二种方法,读者也可试试第一种方法。

② 动态切换当前所选择学院名称左边的书本小图标的打开或关闭图片。如果菜单项索引与当前所单击项的索引相等,则当前所单击学院图片切换为 bookopen.gif 图片; 否则切换为 bookclose.gif 图片。

③ 向后端地址"video/acc"请求视频数据流,视频文件名为 acc。WebFlux 根据该请求路径,读取 static/video 文件夹下的会计学院视频文件 acc.mp4,并推送视频流数据。

2. 后端: SchoolHandler

在 cn.tams.handler 包下新建 SchoolHandler 类:

```
@Component
@RequiredArgsConstructor
public class SchoolHandler {
    @Value("${tams.url.school-video-file}")
    private String videoFile;                              //该变量值为: classpath:/static/video/%s.mp4
    private final @NonNull ResourceLoader resourceLoader;  //注入资源加载器

    public Mono<ServerResponse> videoHandler(ServerRequest serverRequest) {
        String name = serverRequest.pathVariable("name");  //获取路径参数值,例如 acc
        Mono<Resource> mono = Mono.fromSupplier(() ->
                resourceLoader.getResource(String.format(videoFile, name)));
        return ok().contentType(MediaType.valueOf("video/mp4"))
```

```
                        .body(mono, Resource.class);
        }
}
```

videoFile 中的"%s"是格式参数,代表视频文件名,例如 acc。格式化处理形成正确的视频文件名,例如 classpath:/static/video/acc.mp4,由 resourceLoader 加载,然后以流媒体数据形式返回。

13.5.14　学生信息模糊查询

登录后的用户可利用关键字对学生信息进行模糊查询,页面效果如图 13-13 所示。本项功能的实现,换了一种处理思路:前端全部用 render()渲染函数、h()函数来实现,不使用 template 模板。

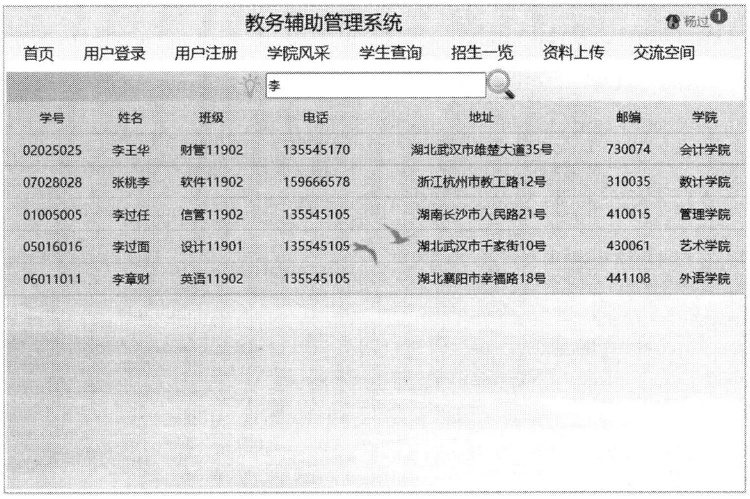

图 13-13　学生信息查询

1. 前端:student-query.component.js

1) StudentQueryComponent 组件

```
const {h, ref, reactive, watch, $ emit} = Vue
const sname = ref('')                              //查询关键字
const surl  = ref('')                              //查询请求地址
const sdata = reactive({
     loading: false,                               //查询是否正在进行标识
     message: null,                                //提示信息
     students: []                                  //查到的学生记录
})
export const StudentQueryComponent = Vue.defineComponent({
     setup() {
          const {buildQueryUi} = QueryMethods       //从函数集中解构出 buildQueryUi()方法
          const {inputStyle} = QueryStyle           //从函数集中解构出 inputStyle 样式
          return () => h('div', {},                 //显示查询结果的层,利用 render()函数渲染返回
```

```
                    [
                        buildQueryUi(inputStyle),                    //构建查询界面
                        h('br'),                                      //换行
                        h(QueryDataComponent, {url: surl.value})      //输出查询结果
                    ])
        }
    })
```

查询请求的地址格式是/query/{key}/{username}，基于此的处理思路是：首先，假定当前登录用户为"杨过"。当其在查询文本框中输入查询关键字，例如"李"，意味着形成了新的查询地址 surl:/query/李/杨过；接着，将 surl 的值传递给 QueryDataComponent 组件的 url；最后，QueryDataComponent 组件内部侦听到 url 值的变化，触发 doQuery()函数，执行具体的查询动作。

代码 h(QueryDataComponent，{url：surl.value})调用 QueryDataComponent 组件查询数据，并将 surl 的值传递给 QueryDataComponent 组件的 url。

2）QueryDataComponent 组件

```
const QueryDataComponent = Vue.defineComponent({
    props: {
        url: {type: String}                          //传递给组件的 url 数据
    },
    setup(props) {
        const {url} = Vue.toRefs(props)
        const {doQuery, setTableData} = QueryMethods
        const {tableStyle, trStyle, redColorStyle} = QueryStyle
        watch(url, newurl => doQuery(newurl))        //侦听 url 值的变化，调用 doQuery()查询

        return () => [                                //利用 render 渲染函数返回查询结果数组
            sdata.loading ? h('span', {}, '正在获取数据,请稍候......') : null,
            h('span', {style: redColorStyle}, sdata.message),
            sdata.students.length > 0 ? h('table', {style: tableStyle}, setTableData(trStyle)) : null
        ]
    }
})
```

3）QueryMethods 函数集

该函数集包括了相应的查询方法：

```
const QueryMethods = (function (exports) {
    const buildQueryUi = (inputStyle) =>          //构建查询界面
        [
            h('img', {src: 'image/note.png'}),
            h('input', {
                type: 'text',
                placeholder: '姓名或班级关键字......',
                value: sname.value,
```

```
                style: inputStyle,
                //由于无法使用v-model进行数据绑定,需要手工绑定数据,这句代码
                //将该文本框的值与sname响应式对象进行数据绑定。
                onchange: event => sname.value = event.target.value
            }),
            h('img', {
                src: 'image/search.png',
                title: '单击查询',
                onClick: setSurl,
            })
        ]
const setSurl = () => {                            //设置查询请求地址
    let name = sname.value.trim()
    if (name.length > 0) {
        surl.value = 'stu/query/' + name
        let loginer = JSON.parse(sessionStorage.getItem('loginer'))
        if (loginer != null)                       //用户已登录,构建新的查询请求地址
            surl.value += '/' + loginer.username
    }
}

const doQuery = async url => {                     //执行查询
    let loginer = sessionStorage.getItem('loginer')
    if (loginer == null) {
        sdata.message = '未登录用户不能查询数据!'
        return
    }
    sdata.students = []                            //清空旧数据
    sdata.loading = true
    sdata.message = ''
    try {
        const {data, status} = await axios({
            method: 'get',
            url: url,
            headers: {
                'Authorization': JSON.parse(loginer).token
            },
            validateStatus: status => status >= 200 && status < 500   //设置响应码范围
        })
        if (status == 401) {
            sdata.message = '登录已过期,需重新登录!'
            tamsPub.clearLoginer()
            return
        }
        data.forEach(s => sdata.students.push(s))
        if (data.length == 0)
            sdata.message = '查无相关数据!'
```

```javascript
        } catch (error) {
            sdata.message = '出错,无法获取数据!' + error
        } finally {
            sdata.loading = false
        }
    }
    //构建表格数据
    const title = ['学号', '姓名', '班级', '电话', '地址', '邮编','学院']
    const setTableData = (trStyle) => {
        let header = []
        title.forEach(t => header.push(h('td', {}, t)))              //表头
        let cell = sdata.students.map(student => {                   //查到的学生对象
            let td = []
            for (let k in student) {                                 //k 指 sno、sname...等属性名
                td.push(h('td', {}, student[k]))                     // student[k]是对应属性的值
            }
            return h('tr', {style: trStyle}, td)
        )
        return [h('tr', {style: trStyle}, header), cell]             //返回有多个单元格数据的行
    }

    exports.buildQueryUi = buildQueryUi
    exports.setSurl = setSurl
    exports.doQuery = doQuery
    exports.setTableData = setTableData
    return exports;
}({}))
```

4) QueryStyle 函数集

该函数集包含了 CSS 样式设置对象:

```javascript
const QueryStyle = (function (exports) {
    const inputStyle = {                        //关键字输入文本框的样式
        width: '220px',
        height: '22px',
        border: '1px solid #f15555'
    }
    const tableStyle = {                        //显示查询结果表格的样式
        width: '100%',
        position: 'relative',
        lineHeight: '30px',
        fontSize: '10px',
        borderCollapse: 'collapse',
        background: 'rgba(229,232,241,0.73)',
        borderBottom: '1px solid #badaff',
        padding: '0px',
```

```javascript
                    textAlign: 'center'
                }
        const trStyle = {                       //表格行<tr>的样式
                    borderBottom: '1px solid #badaff',
                    padding: '0px',
                    fontSize: '13px',
                    textAlign: 'center'
                }
        const redColorStyle = {color: '#ff0000'}
        exports.inputStyle = inputStyle
        exports.tableStyle = tableStyle
        exports.trStyle = trStyle
        exports.redColorStyle = redColorStyle
        return exports;
}({}))
```

2. 后端：Student 实体类

```java
@Data
@NoArgsConstructor
@AllArgsConstructor
public class Student implements Serializable {
        @Id
        private String sno;
        private String sname;
        private String sclass;
        private String tel;
        private String address;
        private String postcode;
        private String cname;            //数据表中无该字段,为了方便增加的属性
}
```

3. 后端：StudentRepository 数据访问接口类

在 cn.tams.repository 包下创建 StudentRepository 类：

```java
@Repository
public interface StudentRepository extends R2dbcRepository<Student, String> {
        //根据姓名或班级关键字模糊查询
        String query = "select s.*,c.cname from student s,school c where " +
                " (sname like '%'||:key||'%' or sclass like '%'||:key||'%') and left(s.sno,2) = c.cno";
        @Query(query)
        Flux<Student> queryStuByKey(String key);
}
```

4. 后端：StudentHandler 业务处理类

在 cn.tams.handler 包下新建 StudentHandler 类：

```
@Component
@RequiredArgsConstructor
public class StudentHandler {
    private final @NonNull StudentRepository sRepository;        //注入 StudentRepository 实例
    //根据关键字进行模糊查询
    public Mono < ServerResponse > queryStuKey(ServerRequest request) {
        Flux < Student > students =
                        sRepository.queryStuByKey(request.pathVariable("key"));
        return ok().body(students, Student.class);
    }
}
```

按照前面路由请求路径的定义,queryStuKey()方法的请求路径为/stu/query/{key}/{username}。

13.5.15　招生数据一览

招生数据一览界面如图 13-14 所示。

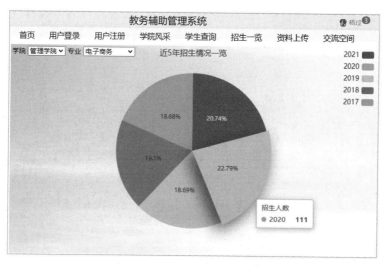

图 13-14　招生数据一览

1. 前端：enroll.component.js

在 resources/static/module 下创建 enroll.component.js 组件：

```
export const EnrollComponent = Vue.defineComponent({
    setup() {
        const {reactive, onMounted, watch} = Vue
        const {chartsOption, setChartsOptionData} = ChartsOptionData
        let sdata = reactive([])                    //图中的序列数据
        const combo = reactive({                    //当前所选的学院和专业
                sindex: null,                       //当前所选的学院,即学院选择框的索引值
                mno: null                           //当前所选的专业,即专业代码 mno
```

```js
                    }
                )
                const schools = reactive([])                      //数组,存放包含学院名、专业名的学院对象
                onMounted(
                    async () => {
                        const {data} = await axios({              //阻塞获取学院名称、专业名称关联数据
                            method: 'get',
                            url: 'stu/query/sm'                   //请求路径,获取学院名、专业名
                        })
                        data.forEach(o => schools.push(
                            {
                                cname: o.school,                  //学院名称
                                major: JSON.parse(o.major)        //专业
                            }))
                        combo.sindex = 0                          //默认选中第一个学院
                    }
                )
                let oldsindex = null                              //学院下拉选择框的旧值
                //侦听下拉选择框学院或专业的变化,只要其中一个发生改变就重置数据
                watch(combo, newCombo => {
                    if (oldsindex != newCombo.sindex) {           //如果是学院发生了变化
                        //需要重置专业选择框,默认显示第一个专业名称
                        combo.mno = Object.keys(schools[newCombo.sindex].major)[0]
                        oldsindex = newCombo.sindex               //保存新的学院索引值
                    }
                    setChartsOptionData(sdata, chartsOption, combo.mno) //设置饼图数据
                })
                return {combo, schools}
            },
            template:
                `<div id = "chart-pie">
                    <div id = "chart-select">学院
                        <select v-model = "combo.sindex">
                            <option v-for = "(school,index) in schools" :value = "index">
                                {{school.cname}}
                            </option>
                        </select>专业
                        <select v-model = "combo.mno">
                            <option v-if = "schools.length > 0"
                                v-for = "(item,key) in schools[combo.sindex].major" :value = "key">
                                {{item}}
                            </option>
                        </select>
                    </div>
                    <div id = "enroll-chart"></div>
                </div>`
        })
```

2. 前端：函数集 ChartsOptionData

```javascript
const ChartsOptionData = (function (exports) {
    const chartsOption = Vue.reactive({
        title: {
            text: '近 5 年招生情况一览',
            left: 'center',
            textStyle: {
                fontSize: '16px',
                fontStyle: 'normal',
                fontWeight: 'normal'
            },
            top: 0
        },
        tooltip: {
            trigger: 'item'
        },
        legend: {                                       //图例样式
            orient: 'vertical',                         //垂直放置
            left: 'right',                              //置于右侧
            textStyle: {                                //文字样式
                fontSize: '14px',
                fontStyle: 'normal',
            },
            top: 0
        },
        series: [
            {
                name: '招生人数',
                type: 'pie',                            //图形类型：饼图
                center: ['50%', '50%'],                 //图形位置
                radius: '75%',                          //饼图半径
                data: null,                             //需要显示的数据
                label: {                                //文本标签样式
                    normal: {
                        show: true,
                        position: 'inner',
                        formatter: '{d}%'
                    }
                },
                emphasis: {                             //强调显示时的样式
                    itemStyle: {
                        shadowBlur: 30,
                        shadowOffsetX: -10,
                        shadowOffsetY: 10,
                        shadowColor: 'rgba(0, 0, 0, 0.5)'
                    }
                }
```

```javascript
                }
            ]
        }
    )
    //构建图形数据
    const setChartsOptionData = (sdata, chartsOption, mno) => {
        axios({
            method: 'get',
            url: 'stu/enroll/' + mno             //查询指定专业代码的招生数据
        }).then(response => {
            sdata = []
            response.data.forEach(enroll => sdata.push(   //查到的招生数据存入 sdata 数组
                {
                    value: enroll.enrollment,    //招生人数
                    name: enroll.nian            //年份
                })
            )
        }).catch(error => console.error(error))
            .then(() => {
                chartsOption.series[0].data = sdata    //具体数据赋值给图形 data 属性
                //删除旧 Echarts 实例,否则导航切换后无法显示图形
                //因为 echarts 每次绘制都需要创建一个新的 Echarts 实例
                let chartDom = document.getElementById('enroll-chart')
                chartDom.removeAttribute('_echarts_instance_')
                let myChart = echarts.init(chartDom)
                myChart.setOption(chartsOption, true)   //重置图形数据
            })
    }
    exports.chartsOption = chartsOption
    exports.setChartsOptionData = setChartsOptionData
    return exports;
}({}))
```

3. 后端:Enroll 实体类

与第 11 章 11.4.1 节代码相同。

4. 后端:EnrollRepository 数据访问接口类

在 cn.tams.repository 包下创建 EnrollRepository 接口类:

```java
@Repository
public interface EnrollRepository extends R2dbcRepository<Enroll, String> {
    //查询近 5 年指定专业代码的招生数据
    String q1 = "select a.nian,b.value as enrollment " +
            "from enroll a,jsonb_each(a.major) b,major c where c.mno = :mno " +
            "and a.nian >= extract(year from now()) - 4 " +
            "and b.key = c.mno group by a.nian,enrollment order by a.nian desc";
    @Query(q1)
```

```
    Flux< Enroll > enroll5ByMno(String mno);
    //查询全部学院名称及对应的专业名称
    String q2 = "select a.cname as school,json_object_agg(b.mno,b.mname)::text as major " +
            "from school a,major b where a.cno = substr(b.mno,1,2) group by a.cname";
    @Query(q2)
    Flux< Enroll > allSchoolMajor();
}
```

5. 后端：StudentHandler 业务处理类

在 StudentHandler 类中加入下面的代码：

```
private final @NonNull EnrollRepository eRepository; //注入 EnrollRepository 实例对象

//查询全部学院名称、专业名称
public Mono< ServerResponse > allSchoolMajor(ServerRequest request) {
    return ok().body(eRepository.allSchoolMajor(), Enroll.class);
}
//查询近5年指定专业代码的招生数据
public Mono< ServerResponse > enroll5Mno(ServerRequest request) {
    Flux< Enroll > enrolls = eRepository.enroll5ByMno(request.pathVariable("mno"));
    return ok().body(enrolls, Enroll.class);
}
```

13.5.16 资料上传

登录用户可上传 50MB 以内的文件，页面效果如图 13-15 所示。

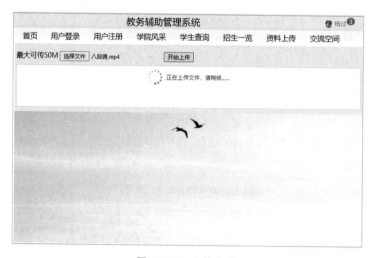

图 13-15 文件上传

1. 前端 file-upload.component.js

在 resources/static/module 下创建 file-upload.component.js 组件：

```javascript
export const FileUploadComponent = Vue.defineComponent({
    setup() {
        const uploder = Vue.reactive({
            isDoing: false,                                    //文件上传是否正在进行的标识
            message: ''                                        //提示信息
        })
        let fileInput = Vue.ref(null)                          //待上传的文件域
        const doUpload = () => {
            uploder.message = ''
            let file = fileInput.value.files[0]                //待上传的文件
            let fData = new FormData()                         //构建表单域数据
            let cancel                                         //取消上传的变量
            const myaxios = axios.create();
            myaxios.interceptors.request.use(                  //添加 Axios 拦截器
                config => {
                    if (typeof file === 'undefined') {
                        cancel('请先选择待上传的文件!')
                        return config
                    }
                    if (file.size > (1024 * 1024 * 50)) {
                        cancel('文件大小超过限制!')
                        return config
                    }
                    let loginer = sessionStorage.getItem('loginer')
                    if (loginer != null) {                     //用户已登录
                        uploder.isDoing = true
                        let user = JSON.parse(loginer);
                        config.url = 'file/up/' + user.username  //请求地址
                        config.headers['Authorization'] = user.token //附加令牌
                        fData.append('uploadUser', user.username)//表单域用户名
                        fData.append('uploadFile', file)       //表单域文件
                    } else
                        cancel('未登录用户,无法上传文件!')
                    return config
                },
                error => Promise.reject(error)                 //回退出错提示
            )
            myaxios({
                method: 'post',
                data: fData,
                validateStatus: status => status >= 200 && status < 500, //正常状态码范围
                headers: {'Content-Type': 'multipart/form-data'}, //设置为多域表单数据
                cancelToken: new axios.CancelToken(c => cancel = c) //取消上传处理
            }).then(response => {
                if (response.status == 200) {
                    uploder.message = file.name + ' 上传成功!'
                    fileInput.value.value = ''                 //上传文件框置空
```

```javascript
                    } else if (response.status == 401) {
                        uploder.message = '登录已过期,需重新登录才能上传!'
                        tamsPub.clearLoginer()
                    } else if (response.status == 417) 
                        uploder.message = '写文件出错,上传失败!'
                }).catch(error => {
                    if (error.request)
                        uploder.message = '服务器未响应,请重试!'
                    else                                //发送请求时出错,例如未登录
                        uploder.message = error.message
                }).then(() => uploder.isDoing = false)    //上传结束,重置状态标识
            }
            return {doUpload, uploder, fileInput}
        },
        template: `
            <div class="file-upload">
                最大可传 50M <input :ref="(input) => fileInput = input" type="file"/>
                <button @click="doUpload">开始上传</button>
                <div class="tooltip">
                    <div v-show="uploder.isDoing">
                        <img src="image/progress.gif" width="50" height="50"/>
                        正在上传文件,请稍候……
                    </div>
                    <div v-show="!uploder.isDoing">{{uploder.message}}</div>
                </div>
            </div>
        `
    })
```

2. 后端：FileHandler

在 cn.tams.handler 包下新建 FileHandler 类：

```java
@Component
public class FileHandler {
    @Value("${tams.file.upload.path}")
    private String uploadPath;

    public Mono<ServerResponse> uploadFile(ServerRequest request) {
        return request.body(BodyExtractors.toParts())    //提取请求中的表单域数据
                .collectList()                            //转换为 List
                .flatMap(parts -> {                       //遍历
                    String username = ((FormFieldPart) parts.get(0)).value();    //文本域
                    FilePart part = (FilePart) parts.get(1);                     //文件域
                    String fileName = part.filename();
                    SimpleDateFormat date =                                       //格式化日期
                            new SimpleDateFormat("yyyyMMddHHmmss");
```

```
                    //存放在项目所在磁盘的tamsfiles文件夹下(需事先手工创建好),例如d:/tamsfiles
                    //文件名格式化处理:日期-用户名-文件名
                    Path path = Paths.get(
                            String.format("%s/%s-%s-%s", uploadPath,
                                    date.format(new Date()), username, fileName));
                    return part.transferTo(path)        //传输文件到目的地
                            .then(ok().build(Mono.empty()))
                            .onErrorResume(e -> ok().bodyValue(e.getMessage()));
                });
    }
}
```

13.5.17 交流空间

在第12章的基础上修改而成。前端变动极少,后端则采用Spring WebFlux的WebSocketHandler技术。由于Spring WebFlux目前不支持SocketJS,所以不再使用。聊天界面效果与图12-6类似。

1. 前端:chat.component.js

前面说过,本章登录令牌的返回发生了细微变化,所以这里也需要稍作改动:

```
setup() {
        const {ref, reactive, onMounted} = Vue
                ...
                message: '连接成功,开始畅论……\n'
            }
        }
        let loginer = null
        let token = null
        let item = sessionStorage.getItem('loginer')
        if (item != null) {
            loginer = JSON.parse(item)
            token = loginer.token
        }
        onMounted(() => {
            if (loginer != null) {
                me.user.username = decodeURIComponent(loginer.username)
                me.user.logo = loginer.logo
                me.user.role = loginer.role
                me.message = ''
                bindListener()
            }
        })
        let sockjs = null
        const bindListener = () => {
            sockjs = new
WebSocket('ws://localhost/schat/' + me.user.username + '/' + token)
            ...
}
```

粗体代码为变化部分，其他代码不变。这里向 WebSocket 传送了用户名、令牌，以便进行有效性验证。

2. 后端：ChatHandler 消息处理类

```java
@Component
@RequiredArgsConstructor
public class ChatHandler implements WebSocketHandler {
    @Value("${chat.topic-name}")
    private String topicName;                                    //Kafka 订阅主题名
    @Value("${chat.logoutMsg}")
    private String logoutMsg;                                    //登录失效提示文字
    private final @NonNull KafkaTemplate kafka;                  //使用 Kafka 模板
    private final @NonNull JWTAssit jwtAssit;
    private static final Map<String, WebSocketSession> map =
                                        new ConcurrentHashMap();

    @Override
    public Mono<Void> handle(WebSocketSession session) {
        map.put(session.getId(), session);                       //保存当前用户
        return session.receive()                                 //接收消息
            .switchMap(wsMessage -> {                            //对身份进行有效性验证
                String payload = wsMessage.getPayloadAsText();   //获取到聊天内容
                String[] uri = session.getHandshakeInfo().getUri().getPath().split("/");
                boolean legal = jwtAssit.verify(uri[uri.length - 1], uri[uri.length - 2]);
                if (!legal) {                                    //身份已失效
                    StringBuffer sb = new StringBuffer(payload);
                    sb.replace(sb.indexOf("message") + 10, sb.length() - 2, logoutMsg);
                    WebSocketMessage msg = session.textMessage(sb.toString());
                    map.remove(session.getId());                 //移除该用户
                    return session.send(Mono.just(msg));         //发送身份过期提示
                }
                return Flux.just(payload);                       //回流聊天文本
            }).flatMap(payload -> {                              //继续处理聊天内容
                kafka.send(topicName, payload);                  //发送给 Kafka
                return Flux.fromIterable(map.values())           //遍历 map 中的每个用户
                    .parallel(Schedulers.DEFAULT_POOL_SIZE)      //多线程并发
                    .runOn(Schedulers.parallel())
                    .concatMap(chater -> {                       //合并消息流
                        WebSocketSession ws = map.get(chater.getId());
                        if (ws != null) {
                            WebSocketMessage msg = ws.textMessage(payload.toString());
                            return ws.send(Mono.just(msg));      //发送给该用户
                        }
                        return Mono.empty();
                    });
            }).then().doFinally(s -> map.remove(session.getId())); //移除已关闭 WebSocket 用户
    }
}
```

代码使用 ConcurrentHashMap 记住所有进入聊天室的用户。当某个用户发送一条消息后,接收该消息,发送到 Kafka。接下来,采用 parallel 多线程并发模式,将该消息合并为待发给其他用户的数据流,然后一起发送。最后,对于已经不在聊天室的用户,则从 ConcurrentHashMap 中移除。

加粗部分的代码,对登录用户的身份进行有效性验证。如果已过期,需要向前端发送身份过期提示。聊天内容 payload 中的字符串类似于下面形式:

```
{"user":{"username":"杨过","logo":"image/dog.png","role":"student"},"message":"你好"}
```

代码使用 replace()方法,将聊天文字"你好"替换成 application.yml 中 logoutMsg 定义的身份失效提示文字"您的登录身份已过期失效,无法向他人发送消息,请重新登录!"。

请比较与第 12 章的差异。为简化起见,用户进入聊天室后,并没有显示最近的 15 条聊天记录,具体处理方法在第 12 章已经介绍过了,读者可以试着完成。

3. 后端:ChatConfig 配置类

```java
@Configuration
@RequiredArgsConstructor
public class ChatConfig  {
    @Value("${chat.ws-endpoint}")
    private String wsEndPoint;
    private final @NonNull ChatHandler sChat;
    //注册映射路径"/schat"
    @Bean
    public HandlerMapping handlerMapping(){
        Map<String, ChatHandler> handlerMap = new HashMap<>();
        handlerMap.put(wsEndPoint, sChat);
        return new SimpleUrlHandlerMapping(handlerMap, 1);
    }
}
```

13.6　打包发布

打包前,请先关闭测试模式:单击图 13-16 圆圈中的按钮,切换为 Skip Tests 模式。再单击 Maven 中的 package 选项(见图 13-16),执行打包过程。不久会提示打包完成,在项目的目标文件夹下将生成 jar 文件,即 tams\target\tams-0.0.1-SNAPSHOT.jar。

现在,可退出 IntelliJ IDEA。不妨将 tams-0.0.1-SNAPSHOT.jar 复制到 D 盘,并在 D 盘创建好 tams 运行时存放上传文件的文件夹 tamsfiles。进入 DOS 命令提示符,输入命令:

```
java -jar tams-0.0.1-SNAPSHOT.jar
```

图 13-16　Maven 打包

将启动 Netty 服务器并加载 tams 教务辅助管理系统，如图 13-17 所示。现在，在浏览器中输入 http://localhost/，将会看到熟悉的 TAMS 主页画面（图 13-1）。

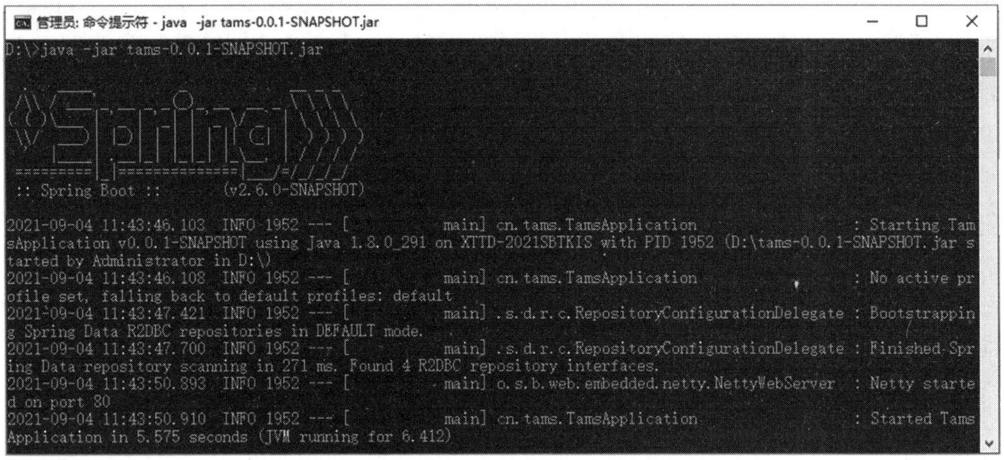

图 13-17　启动 tams 应用

图书资源支持

感谢您一直以来对清华版图书的支持和爱护。为了配合本书的使用,本书提供配套的资源,有需求的读者请扫描下方的"书圈"微信公众号二维码,在图书专区下载,也可以拨打电话或发送电子邮件咨询。

如果您在使用本书的过程中遇到了什么问题,或者有相关图书出版计划,也请您发邮件告诉我们,以便我们更好地为您服务。

我们的联系方式:

地 址:北京市海淀区双清路学研大厦 A 座 714

邮 编:100084

电 话:010-83470236 010-83470237

客服邮箱:2301891038@qq.com

QQ:2301891038(请写明您的单位和姓名)

资源下载: 关注公众号"书圈"下载配套资源。

资源下载、样书申请

书 圈

获取最新书目

观看课程直播